Reason in the City of Difference

D1143791

In the modernist city rationality ruled and subsumed difference in a logic of identity. In the postmodern city, reason is abandoned for an endless play of difference. *Reason in the City of Difference* poses an alternative to these extremes by drawing on classical American philosophical pragmatism (and its contemporary developments in feminism and the philosophy of communication) to explore the possibilities of a strengthening and deepening of reason in the contemporary city. This is a transactional rationality based on communication, rather than cognition, involving bodies as much as minds, and non-discursive, as well as discursive competences. It is a rationality that emerges out of difference and from within the city, rather than over and above it.

Using pragmatist philosophy and a range of suggestive examples of urban scholarship, *Reason in the City of Difference* offers a new, alternative reading of the city.

Gary Bridge is Senior Lecturer at the Centre for Urban Studies, School for Policy Studies, University of Bristol.

Questioning Cities
Edited by Gary Bridge, *University of Bristol, UK* and
Sophie Watson, *The Open University, UK*

The 'Questioning Cities' series brings together an unusual mix of urban scholars under the title. Rather than taking a broadly economic approach, planning approach or more socio-cultural approach, it aims to include titles from a multi-disciplinary field of those interested in critical urban analysis. The series thus includes authors who draw on contemporary social, urban and critical theory to explore different aspects of the city. It is not therefore a series made up of books which are largely case studies of different cities and predominantly descriptive. It seeks instead to extend current debates, in most cases through excellent empirical work, and to develop sophisticated under-standings of the city from a number of disciplines including geography, sociology, politics, planning, cultural studies, philosophy and literature. The series also aims to be thoroughly international where possible, to be innovative, to surprise, and to challenge received wisdom in urban studies. Overall it will encourage a multi-disciplinary and international dialogue, always bearing in mind that simple description or empirical observation which is not located within a broader theoretical framework would not – for this series at least – be enough.

Published

Global Metropolitan
John Rennie Short

Reason in the City of Difference
Gary Bridge

Forthcoming titles:

Ordinary Cities
Between modernity and development
Jenny Robinson

Reconsidering Public Space
The multicultural practices of everyday life
Sophie Watson

Reason in the City of Difference

Pragmatism, communicative action and contemporary urbanism

Gary Bridge

Routledge
Taylor & Francis Group

LONDON AND NEW YORK

First published 2005
by Routledge
2 Park Square, Milton Park, Abingdon, Oxon, OX14 4RN

Simultaneously published in the USA and Canada
by Routledge
270 Madison Ave, New York, NY 10016

Routledge is an imprint of the Taylor & Francis Group

© 2005 Gary Bridge

Typeset in Times and Bauhaus by Exe Valley Dataset Ltd, Exeter
Printed and bound in Great Britain by The Cromwell Press, Trowbridge, Wiltshire

British Library Cataloguing in Publication Data
A catalogue record for this book is available
from the British Library

Library of Congress Cataloging in Publication Data
A catalog record for this book has been requested

ISBN 0–415–28766–9 (hbk)
ISBN 0–415–28767–7 (pbk)

For Deborah

Contents

Preface

Contemporary urban theory seems divided between homogenizing circulations of power and increasingly complex, social relations of difference. In recent social theory rationality is cast as the agent of power, be it economic, disciplinary or discursive. What I do in this book is suggest how the city of difference may play host to an alternative form of rationality – transactional rationality – which emerges from the communicative potentials of a more networked and distantiated urban space.

An appreciation of transactional rationality comes from some recent developments in a philosophy that influenced some of the earliest systematic theorizations of the modern city through the Chicago School – that is, American pragmatism. Contemporary feminist pragmatism and pragmatist theories of communication draw on the insights of the classical pragmatists, and especially the work of John Dewey. This work includes a broad idea of communicative action and rationality (discursive and non-discursive, aesthetic as well as instrumental, speculative as well as conventional) involving bodies as much as minds (or rather body-minds) that has emancipatory potential (as well as reproducing disciplinary norms). In the structure of the book I explore the traditional and non-traditional sites of rationality in the city: On the body (Chapter 2); On the street (Chapter 3); In the community (Chapter 4); In the public realm (Chapter 5); At work and home in the urban economy (Chapter 6); In city hall (Chapter 7); and conclude with Cosmopolitan reason and the global city (Chapter 8). In all these sites I suggest how transactional rationality opens up alternative spaces of communication involving ideas of bodying, communicative excess, hybridity, emergent publics, pragmatic planning as argumentation, and an idea of cosmopolitanism as situational. All these realms show how rationality is implicated within difference in the city.

Throughout the book I draw on a range of historical and contemporary examples of urban scholarship on western cities that I think can also be deployed to illustrate the spaces of transactional rationality. From the work of the women professionals in the early settlement houses, to an analysis of civic clubs and social centres in US cities, to the emergence of gay identity in New York to Mardi Gras in Sydney and New Orleans to my own work on gentrification in Sydney and London, I pursue these and other examples in some depth to try to convey the importance of the urban 'situation' in understanding rationality and difference.

Reason in the City of Difference makes a case for the contemporary relevance of philosophical pragmatism in theorizing the city. It posits not the modernist city where rationality is sameness, nor the postmodern city where rationality is abandoned to an endless play of difference. Rather it suggests how, in myriad spaces of communication in the city, difference can still be brought into dialogue.

Acknowledgements

I am grateful to the five anonymous referees of the original proposal for this book. Their insights were instrumental in taking me in interesting and challenging directions. Previous versions of chapters 3, 5 and 8 were given respectively at the Institute of British Geographers Conference (Plymouth 2001), the ESRC seminar series 'Transforming Cities' (London 2002) and the Association of American Geographers Conference (New Orleans 2003). My thanks to all who hosted these sessions and to the participants whose feedback helped me on the way. Particular thanks to Bob Catterall, who originally took up the book proposal, and Andrew Mould and Anna Somerville at Routledge, who have been extremely helpful and supportive. Sophie Watson I can't thank enough. Her advice and encouragement have been invaluable: a great colleague and true friend.

1 Reason in the city of difference

INTRODUCTION

The city has always been the home of reason. From early Greek civilization the city was the polis, the location of political democracy. The agora was the site of open and enquiring debate amongst rational citizens. Yet that space of reason also excluded difference: women and slaves did not count as rational citizens. This association of reason with exclusivity and the activities of an elite persisted into the modern era, encoded in the urban Master Plan and efficient bureaucracy. In the past century the idea of reason has been assailed from all directions – through a turn to the body, to language, to culture, to the unconscious. The way that difference is registered across all these realms is now of primary philosophical and social concern. That the city should be the home to difference is an aspiration in metaphor as well as practical politics – where cities pre-eminently include the claims of multicultural and polyvalent identities. Surely it is time to banish reason, with all its exclusivities and homogenizations, from the city, and to let difference in?

THE CITY AFTER POSTMODERNISM

In his paper heralding the era of postmodernity Frederick Jameson (1984) pointed to the Bonaventure Hotel, a structure of endlessly curved mirror glass that refracted the city around it but gave no sense of its own interiority, as iconic of the new age. The postmodern turn in urban studies denied that explanations of the city could rest on ultimate foundations of knowledge and opened up the space for a cultural politics of identity (sexuality, gender and ethnicity as well as class). It treated space as multiple and emergent and critiqued the desire to locate or fix space, in the same way poststructuralist theory sought to evade the fixing of meaning in the word. And yet its anti-foundationalism and emphasis on difference ultimately meant that the postmodern city was seen as all surface, an endless play of space and difference, an unmappable space, a posthuman environment in which human activities are just one small part of an overall assemblage of emergent effects that involve also non-human biological actants, machines and texts.

In response to the claims of postmodernism, and whilst acknowledging the significance of its challenge, some urbanists claimed an interior, an identity, a stability for the city, based on certain urban orderings – be it the logic of capital, discourses of power and surveillance, or rational planning. So we have conceptions of urban space that see its coherence through the circulation of power or its pluralization through displacement and difference. These divisions can be seen in distinctions between system and lifeworld (Habermas), abstract space and lived space (Lefebvre), strategy and tactic (de Certeau), disciplinary space and heterotopias (Foucault), the space of flows and everyday life, the public and the private realm.

What I do in this book is suggest that there is another set of responses to the challenge of the postmodern city. It is a tradition of thought that has resonance with (and indeed was partly implicated in) postmodernism but it also has an earlier connection to the city: that of philosophical pragmatism. Over the last two decades there has been a rapid expansion of interest and debate around pragmatism (Rorty 1982; West 1993; Dickstein 1998; Stuhr 2003; Joas 1993). Recent developments in pragmatism have the spirit of postmodernism in the recognition of difference, but also capture some of the significance of communication and discourse after the linguistic turn in philosophy. The purpose of this book is to suggest how a certain reading of pragmatism gives us an understanding of a rationality that can live with difference, that in some senses comes out of difference and the nature of contemporary urban space. It is an understanding that does not dichotomize urban space into the instrumental or the communicative, the system and the lifeworld, abstract and lived. Rather it sees their situational interweaving in spaces of communication that are non-discursive as well as discursive. It is performative but also full of articulation and interpretation. The diversity of space-times of communication in the city is where what I call 'transactional rationality' is made manifest.

I think this leaves us with a different kind of city, one in which the space of power is not separate from the pluralization of difference. It is neither the postmodern city of pastiche – of separate and incommensurable social worlds that are endlessly emerging – nor is it the modernist city held together by singular orderings of capital, discourse or reason. It is instead an under-standing of the city that is full of communicative difference but in which there can be an evaluation of claims across dissensus and difference, the resources for which come from the range of transactions rather than any appeals to transcendence. It is a city in which rationality is an attempt to respect difference through discursive and non-discursive argument and interpretation that goes on not in some singular public realm but in the myriad space-times of communication that exist between communities, within communities and within the ongoing project of the self.

All these themes are central to current understandings of the directions of urbanism, and all, I argue, provide the possible basis for a strengthening and deepening of reason in the city. But this is a reason very different from its forebear in modernity. Rather than being confined solely to mind, it involves body-mind as a form of intelligence. Rather than ignoring habitual action, habit provides its motive force. Rather than being based on an autonomous mind of the individual, it is built out of the social relations that make individuality possible. Rather than being confined to linguistic communication it also involves non-linguistic competences. Rather than being cool-headed and detached from emotion, emotion provides its focus. Rather than requiring coherence and self-presence, it is prompted by absences and delays. All in all it is a capacity that has always existed in the city but too often has not been able to be expressed. It is a rationality that works from within the city rather than over and above it.

RETHINKING PRAGMATISM – REVISITING CHICAGO

Philosophical pragmatism has a long-standing connection to the city. It was the work of the classical American pragmatists (Charles Sanders Pierce 1839–1914, William James 1842–1910, John Dewey 1859–1952 and George Herbert Mead 1863–1931) that was *the* major intellectual influence on the first sustained effort at urban theory: the Chicago School of urban ecology that came out of the Department of Sociology at the University of Chicago in the early decades of the twentieth century. There are as many pragmatisms as postmodernisms at present and even classical pragmatism was host to a range of positions from the critical realist pragmatism of Charles Pierce that assumed a world out there of which humans had fallible knowledge, through to the perspectivism of John Dewey, in which knowledge and the world was an outcome of action in different socio-cultural settings.

Despite differences there are a number of characteristics that broadly define pragmatism. One is the judgement of knowledge by its practical consequences rather than theoretical coherence. This is related to its idea of praxis: that knowledge is something that must be practically acted out. It must be tested by its consequences, rather than its a priori logical propositions. Logic is more like a process of inquiry, rather than a set of abstract propositions. 'Reality' is seen as unpredictable and emergent. Thus knowledge is fallible and always provisional.

One of the things Chicago School urbanists such as Park, Burgess and McKenzie took from classical pragmatism was its emphasis on the organic nature of life. Whereas positivism and mechanics were taking social science in the direction of causality and linearity and the borrowing of models from the natural sciences, classical pragmatists stressed the organic web of life as the basis of understanding social relations. For the Chicago School the city was seen as an ecosystem (and more specifically a plant ecosystem) in which different social groups vied for space and survival (Park *et al.* 1925). The growing metropolis of Chicago in the 1920s presented a challenging environment to which new immigrant communities had to adapt. Each community adjusted in its own way by developing its own conventions of activity in the city. But often these adaptations were undermined by the power of economic forces to disrupt social relations and re-sort people not in terms of their communities of ethnicity but rather in terms of the community of money. People were sorted in urban space according to economic status rather than community heritage. These theorizations were based on the detailed ethnographies of different community 'situations' that constituted the work of the early Chicago School and still define this tradition today (Fine 1995).

There is, I think, an understanding of rationality at the heart of the work of the Chicago School which has not been so remarked upon in urban studies. For Park (1926) rationality was the ability of an individual to be understood in public, which meant by her or his 'moral community'. It was an idea of expression and acceptance in public. So, rationality was a form of communication, and one that is limited by one's cultural community. Park argued that economic rationality undermines community, or communicative rationality resulting in social disorganization.

The idea of rationality as a relationship to one's community is advanced strongly in the 1930s onwards by George Herbert Mead. Developing Dewey's anti-Cartesian idea of the development of mind as something socially shared, rather than being an individual attribute, Mead takes forward an understanding of the role of communication in the formation of self and community (Mead 1934). Mind is formed when gestures shared between people (or animals) come to mean the same thing for the two participants, such that they can anticipate and influence each other's behaviour. This ability to share gestures in mutual anticipation *is* rationality in action. In humans it develops (over human history and in the social development of each individual) into shared symbols, the most sophisticated and resourceful of which, is language. 'Mind' is an attribute of language. Mind as a social attribute is held together by rationality as a form of communication. For Mead rationality relied on the 'me' or 'generalized other', the ability of the individual to anticipate the response to an action by the generalized other of the community as a whole. This contrasts with the 'I', the unpredictable side of the self, the improviser, the historical agent. Mead gave most emphasis to the 'me' in ongoing social relations.

To Mead's emphasis on the 'me' of community Herbert Blumer (1969) stressed the symbolic forms of communication at the expense of the non-symbolic forms. This gave rise to a whole field of studies, symbolic interactionism, the analysis of shared symbols in social interaction out of which meaningful social worlds are built. This flowed out of Chicago School research and continues as a strong research tradition today (Fine 1995; Plummer 1997). So we have a double emphasis: on language as the significant set of symbols in communication, and, the point I wish to emphasise, on rationality as a facilitator of communication, but strictly within community. This is a continuation of Park's thinking about the separate social worlds of the city, where the city's size is able to support a critical mass and separate and diverse ways of life (witness Gans' (1962) 'urban villagers'; Suttles 1968; Peach 1975).

The difficulty with community ethnographies that typified the Chicago School was that they were based on microanalysis and took community norms at face value. Such studies were seen as too parochial to capture these larger circulations of power (Flanagan 1993). This was evident with the strengthening of functionalism and systems theory in urban studies but especially true after the advent of Marxist urban studies with the publication of David Harvey's *Social Justice and the City* (1973). Community-based analysis had no way of explaining the logic of capital accumulation and the urban process in capitalism (Harvey 1978). There was a wider rationality of the system at work that often undermined the community and communicative rationality.

The way that system rationality undermines communicative rationality is the theme taken up by Jurgen Habermas in his reconstruction of social theory (Habermas 1984, 1987). Habermas's work connects to urban studies to some extent, especially in the work of Richard Sennett (1974, 2000) in his discussion of the public sphere (which I consider in Chapter 5), and through the 'deliberative turn' in planning theory (discussed in Chapter 7). Habermas combines the heritage of Marxism and critical theory in his understanding of 'the system', with an understanding of communication in the lifeworld that is based in pragmatism, and especially the work of Mead.

Habermas's work suggests how communication, rather than simply being held within community boundaries by the binding force of communicative rationality, is held in check from without communities by the instrumental power of the system (in the form of the economic rationality of the capitalist system and the instrumental rationality of modern state bureaucracy). This is what Habermas calls the colonization, or elsewhere the provincialization, of the lifeworld by the system. Communication is distorted by capitalist ideology and removed from a communicative context by the steering media of power and money. For Habermas this is a split within rationality itself, between the instrumental rationality of the system (economic rationality and bureaucratic utilitarianism) and the communicative rationality of the lifeworld. Following a division introduced by speech act theory (Austin 1962; Searle 1969) Habermas makes this separation based on the functions of language oriented to action (instrumental) and language oriented to mutual understanding without a press to action (communicative).

PRAGMATISM AFTER POSTMODERNISM

Habermas's view of communicative action is limited to language and to a representative view of language at that. Postmodernism marked a crisis of representation and, along with poststructuralist interventions, questioned the idea that language represents reality. Habermas also has a procedural idea of rationality, one that rests on a 'universal pragmatics' in which implicit validity claims underlie all language use. The contents of the validity claims relate to people's ideas of the objective world, their social norms and their subjective worlds. In order to defend these worldviews in debate people must examine the basis of their validity – in terms of objective truth, social legitimacy and subjective sincerity. This reflection on one's own assumptions as well as the examination of other people' objective, social and subjective worlds is deliberative and communicatively rational. However the rational 'good reasons or grounds' on which participants defend their validity claims still stresses cognition and formal communication, despite Habermas's wish to move away from a philosophy of consciousness towards a philosophy of language and intersubjectivity.

In contrast there is a view of communicative action that is much broader, including non-discursive performativity, as well as discursive communication. This approach to communication involves bodies and gestures, as well as speech and thought. It suggests that there might be all kinds of uncontrollable effects (or excess) around communication. Communicative action is fraught with inconsistencies, slippages and misunderstandings. Performativity, slips and excess in communication can be as much a resource for social transformation as the more controlled communication towards consensus, on which Habermas focused. This is the point made by Judith Butler in her work on gender norms and communication (Butler 1993, 1997). But whereas Butler looks to the effects of body excess, or communicative slips as a sort of universal quality of communication, the pragmatist approach suggests how these effects are qualities of the particular situation and communicative transaction, rather than universal qualities per se (Sullivan 2001). The pragmatist approach also suggests how communicative action constitutes situations, rather than representing them, or being contained by them. It sees dissensus being as much part of the communicative situation as consensus, and speculation being as significant as conformity.

These arguments for a more full-bodied view of communicative action (and communicative rationality) come from a group of pragmatist philosophers working in the philosophy of communication and feminism, and especially in the work of Lenore Langsdorf, Shannon Sullivan, Sandra Rosenthal and Charlene Haddock Siegfried. This deepening and broadening of the realms of communicative action is a contemporary renewal of the work of John Dewey and takes in many of the developments in contemporary philosophy after postmodernism and after the linguistic turn. This 'communicative turn' develops Dewey's dissolution of the Cartesian distinctions between mind and body (to body-mind) and suggests the mutual implication of instrumental and communicative rationality, argument and aesthetics, and system and lifeworld. From this perspective communicative action is implicated in systems of dispersal of power (in a Foucauldian sense) as well as being in resistance to power. Resistance is at the heart of power, rather than being provincialised and separate from it. Dissensus exists within as well as between communities, and indeed within argument and voice, as well as between them.

THE CITY OF REASON AND DIFFERENCE

A broadening and deepening of the idea of communicative action in the way suggested I think gets to the heart of contemporary debates about the city and urban space. There are the conditioning forces of tradition and the circulation of power/discourse. These 'conserving' forces Dewey (1922) called 'habit', a form of productive disposition. But tradition and power are not impervious. Non-discursive performativity and communicative 'excess' suggest that rationality is not limited to a form of practical reason within social boundaries. Rationality is not just confined to community, but overspills its limits. Individuals are not necessarily confined to community but increasingly operate in networks of overlapping communities – with different ties and pulls that sometimes rub against each other – a clash of habits. The city gives the chance of diverse connections as well as enclosure within enclaves: there is hybridity as well as singularity. The reproduction of structures of identity and power relies on everyday performativity of speech acts and non-discursive body communications. This approach is sensitive to the whole communicative repertoire that includes slips and give-aways that reveal other subconscious states. There are hints, innuendos, swarms of unratified messages that transact in the city. There are heart-stopping, world-disclosing moments that change the course of action and the life-course.

As those advocating the postmodern city argue, there are more voices and a greater diversity of connection, a proliferation of difference. But this might result in a city of disconnection, the city as a patchwork of difference between which it is impossible to communicate, or where any attempt to do so is an act of domination. Indeed some contemporary pragmatists think this way. Richard Rorty (2000) believes that rationality is a form of loyalty to community and that it is not possible to bridge different spheres of rationality. Ideas are just elements of conversation but are not decisive. What there should be is a strict separation of the public and the private realm, with the public delimited by basic tolerance and a minimal commonality in the mutual avoidance of pain and harm (Rorty 1988, 1991).

What I suggest in this book is that rationality cannot be confined to community in this way. There is a release of forms of 'rational' communication within community that leak out beyond it, between bodies (Chapter 2), in brief encounters on the street (Chapter 3), in the community (Chapter 4), in the public realm (Chapter 5), in the economy (Chapter 6) and planning more widely (Chapter 7). What I am emphasizing is that these communicative actions are also a resource for rationality. They are part of the speculative action that occurs when different spheres of rationality collide. It is the rationality that includes Dewey's 'I' of historical agency, as well as the 'me' of the generalized expectations of community. This form of argumentation and interpretation takes place within community, as well as between communities.

Rationality is the attempt to meliorate all the conflicting (pre-cognitive) traditions, body interactions, emotions and cognitive thought, both within community and the self. Argumentation is part of this process; However, it does not rely on universal ways of validating claims but on ways that emerge from the resources of the participants in the transaction. Whereas it was condemned as a force of rationalization in modernity, postmodernism sought to banish rationality from the city altogether. I suggest another fate for rationality since postmodernist claims for the recognition of difference: not its banishment from the city as the heinous logic of identity, but rather its release from the confines of community but without then having to rely on the claims of universality. Rational argumentation continues to exist in the plethora of constitutive, communicative engagements, the babble and buzz of the city.

THE SPACE-TIMES OF THE CITY

How does the communicative array suggested by these arguments relate to space and times of the city? There is a good deal to suggest that they might be peripheral to the inner fashioning of urban space. The classical models of urban form were predicated on an idea that centrality was a constitutive feature of the urban. Nowadays the niching of production and the segregation of consumption are leading to a radically decentralized urban form consisting of bundles of transactional activities (Scott 1989). Added to this is the decline of the importance of the city as a central place for social relations, a public space (Sennett 1974, 2000).

Decentralization and distantiation via communication technologies can mean that much of the activity in the city is more and more automatic (Amin and Thrift 2002). The everyday experience of the city is full of imperatives that, benignly or malignly, make up much of human life (Lingis 1998). The flow of pedestrians on the street, traffic signals, automatic doors, swipe cards and software to help organize thought – human activity is more and more infused with technology, to which much of the responsibility for ongoing activity has devolved.

Distantiation and automation relate to the fact that the properties that characterise cities are increasingly emergent and relational. Cities are no longer built on fixed assets as sites of the manufacture of raw materials or the exchange of goods but rather rely on relational assets such as place marketing and inter-urban competition (Amin 2000). It is also suggested that cities no longer comprise fixed social identities based on predictable labour markets – the prevailing environment is more risky and flexible. In terms of their physical and social characteristics the compelling metaphor for cities of late has been as constellations of emergent networks in spaces of flows (Castells 1996, 1997, 1998).

Automatic mediations, stretched out and less predictable relations, relational assets many of them based on image management – the city is more and more coming to consist of signs and surface forms or simulacra (Baudrillard 1981), perfect copies that dissolve the distinction between the authentic and the derivative. The city of surfaces was the privileged icon that heralded postmodernity. Los Angeles has been seen as emblematic of the postmodern urban condition (Dear 2000) or the postmetropolis (Soja 1997). For others these changes signalled the latest phase of capitalist accumulation (Harvey 1989) or a more liquid modernity (Bauman 2000).

This changing idea of urbanism can be seen as a move from the rational to the post-rational city. If cities are more and more decentralized and distantiated, emergent and networked, full of automatic activities and surface manifestations this works against the rational city in a number of ways. Postmodern planners suggest that the decentred and distantiated city works against a planning rationality that seeks to conceive of the city as a whole with coherent, specialist sub-districts that contribute to the overall efficiency of the urban system (Beauregard 1989; Dear 2000; Sandercock 1998). If cities are emergent and networked this works against an idea of urban citizens as having given preferences from stable identities which come into conflict and are argued over in the urban political arena (Laclau and Mouffe 1985). It also denies fixed ends of activity to which rationality provides the most expeditious means. If many of the activities that go on in cities are automatic and coordinated increasingly by machines, then the idea of rational choice and deliberation and the scope for influence of such human rationality is much reduced (Amin and Thrift 2002). And if the city is increasingly about surface and serendipity and an endless play of difference, this denies any ontological depth, which the idea of human reason has claimed historically.

Many urbanists have been happy to bid farewell to reason (Soja 1996; Dear 2000; Gibson and Watson 1995). It has been all too implicated in certain forms of construction of the public realm, in rational planning regimes that focus on means–ends efficiency and the rationality of the market, in the instrumentalization of bodies and the limiting of communication to instrumental action.

In this book I suggest that distantiation and decentralisation are the very conditions in which a transactional rationality might operate. The prompts for this reconstruction of rationality are some emergent and intensifying characteristics of, and concerns with, the contemporary city and especially with city as a site of difference (Fincher and Jacobs 1998). These character-istics are all to do with the nature of urban space and the way that it is relationally constituted with objects and humans.

Cities are assemblages of overlapping and distantiated communication networks; these networks involve acting objects as well as humans. As well as being hubs of mediated relations, the heterogeneity of cities still maintains the capacity for direct encounters and unpredictable situations. Bodies meet in the city. These encounters involve linguistic and non-linguistic communic-ation. The encounter with the other involves this broader set of communicative competences but also connects to the wider net of distantiated relationships. Encounters between different groups and interests in the city are often also emotionally charged and a challenge for urban politics is to cope with these conflicts. Cities are also the sites of wider instrumentalities, especially those of work. Here the ideas of Dewey suggest a broader understanding of instrumentalism to include technical-experimental activity (Buchholz and Rosenthal 2000). It might be that there are inherent tensions in the use of technology and other implements – over whether they are confined to the narrow instrumentalism of the capitalist economy, or whether they might be used to broaden contact and communication to serve more emancipatory goals.

Transactional rationality has a very different relationship to difference. Whereas traditional rationality sought to absorb difference within universality of content or procedure, a 'logic of identity', this rationality recognizes the socialized construction of self, or rather selves, and rationality of as the melioration selves. Difference is a part of identity. The self is heterogeneous. This rationality does not converge on teleological goals but has ends-in-view. It is not convergent instrumentalism, rather instrumental purposes involve contacts that are quasi-experimental and that can alter what gets done: instrumental purposes are more open to difference. Delays, absences and novelty enhance rational capacities. The breaks in communication and mediation of contact start to shade into the space of difference, one that is neither here nor there, neither self nor other.

This approach to *rationality* suggests that Habermas's distinction between system and lifeworld, like the distinctions between micro and macro or structure and agency, are overly dichotomized. One clear way this is registered, discussed throughout the book, is that even the distinction between instrumental (or strategic) rationality and communicative rationality starts to break down very quickly. Communicative gestures often have instrumental goals. An understanding of rational choice theory and game theory indicates how rapidly even narrowly self-interested action takes on social forms – and just how difficult it is to achieve coordination as a result. Equally, by taking a *relational* approach to the social and the individual, micro–macro distinctions start to disappear. Rather than overbearing structures determining action, power is present (and absent) in social/technical networks and in the constitution of self. Thirdly, by focusing on the production of *urban space* we get an idea of the kaleidoscopic nature of phenomena, the whole-in-part, and how miniature movements can capture much larger urban rhythms.

Transactional rationality involves both space-time distantiations and compressions in urban space. These space-times are constituted by a range of communicative spheres comprising community 'habits', transactions between bodies, and a range of discursive and non-discursive interactions (from the instrumental to the aesthetic) of varying intensities and durations. Nevertheless it is productive to consider the city in terms of the locations where different forms of rationality are assumed to dominate. The book is structured around these key locations and their dominant rationalities as well as alternative transactional rationalities that might exist in these spaces.

THE SPACES OF RATIONALITY IN THE CITY

The city is a collection of bodies. Bodies might themselves be rationalized by economic forces and instrumental bureaucracies as Foucault (1977, 1978) has shown so well. Alternatively bodies are seen as sites of resistance to rationality (Lefebvre 1991). They are non-cognitive, habitual and sensory, providing alternative orientations in urban space. Bodies are produced in urban space but they also constitute that space. In Chapter 2 'On the body' I argue that the body is neither wholly submerged nor in resistance to rationality but rather there is a mind-body interaction that is balanced differently in different spaces of the city. It develops Dewey's idea of body-mind and a more active notion of habit to suggest how body-mind operates in usual and novel situations. It takes Sullivan's (2001) reading of Dewey to suggest how one form of communication is 'transactional bodying' that involves 'had' knowledge as a form of ongoing intelligence. This includes elements of habit and disposition but also performativity and communicative excess. It offers a pragmatic, situational, alternative to Judith Butler's (1993, 1997) ideas of the reproduction of norms and social transformation. Rationality here is a form of embodied rapport. Transactional bodying is explored in a number of urban situations, from situationist derives, clubbing and raves, flash mobs and glamours.

Simmel (1950) famously noted how the excess of stimulation in the modern city leads to a form of urban rationality in social interaction based on mutual indifference between strangers. The size and heterogeneity of the modern city makes it impossible for strangers to meet openly and emotionally. In Chapter 3, 'On the street' I argue that even in this most pared down communicative situation there is an excess of messages that start to expand communication. Narrow urban social relations guided by instrumental rationality are in fact structured in the same way as much richer forms of communication. This is explored via Goffman's work on interaction, in ethnomethodology, and the ethnographic work of the later Chicago School. Street communication includes non-verbal body performances. These are forms of communication that cannot be represented in language but are performed by communicating bodies. Dewey accepts a strong performative element of the 'I' (historical agent) and its relation to Mead's (1934) idea of the 'generalized other' (the 'me' of community expectation) and this suggests the capacity for innovation and creativity in interaction. It gives the potential for an opening up to others rather than the stark limit of communication that Simmel originally observed as urban rationality. The chapter explores how the range of communications on the street can lead to a closing down or opening out of interaction. De Certeau's (1984) pedestrian rhetoric suggests how ways of walking and talking can improvise on the structure of communication. I compare Mardi Gras in New Orleans (Shrum and Kilburn 1996; Jankowiak and White 1999) and Sydney (Bruce *et al.* 1997) to suggest how communication at carnival can lead to a reproduction of social norms (in New Orleans) and a transgression and challenging of those norms (in Sydney). I also explore examples of the interruption of pedestrian flow to reconfigure social relations (in a Melbourne artspace – Rossiter and Gibson 2000) and to make a political statement (the street politics of Jackie Smith – Jones 2000).

Ferdinand Tonnies' (2001) distinction between gemeinschaft and gesellschaft influenced questions about whether social relations in the city were community-like or based on specialized association. The point I make in Chapter 4 'In the community' is that the distinction between gemeinschaft and gesellschaft in as much about division between different ideas of rationality as it is about different forms of settlement and social interaction. There is the tacit and intuitive rationality of gemeinschaft and the contractual and abstract rationality of gesellschaft. This chapter explores the relationship between rationality and community. In contrast to Tonnies' assumptions of communicative rationality within community and instrumental rationality without, I argue that rationality is transactional both between and within communities and not always in conformity to community norms. This is explored in the context of the speculative rationality of the Hull House settlement in Chicago (through Addams' work 1968, 1910), and in the situational and transactional forging of gay identities in Chauncey's (1994) analysis of the rise of gay New York. I go on to discuss the tensions between community and hybridity in the city.

Chapter 5 'In the public realm' explores Habermas's (1984, 1987) communicative rationality and Sennett's (1974) ideas of performativity in public as part of a wider interpretation of communication, rationality and the public realm of cities. Using Fraser's (1992) and Young's (1990, 2000) work it looks to the edges of communication as significant in setting the tone for public debate. I include non-discursive and speculative behaviour as part of public deliberation and a rationality in public that involves improvisation as well as conformity. These themes are explored using Mattson's (1998) investigation of public discussion in the civic clubs of the Progressive Era as a detailed case study. I also look at contemporary spaces of communication that might open up the public realm, through music for example, as a form of transactional rationality. The possibilities of a fuller public realm are in part the experience of being the city and the way that selves and others are constituted in interaction.

Economic rationality is the dominant form considered in Chapter 6, 'At work and home in the urban economy'. I suggest how, following Veblen (1899), economic rationality can be distorted by the forms of communication that aim to convey status. The conspicuous consumption of elite gentrification is explored via my research in Sydney and London. I pursue the idea that economic processes are often embedded in particular socio-cultural situations, of which what I call the 'gentrification premium' is a prime example. More recently the city has become a key arena for understanding the interrelationships between the economic and the cultural. The embeddedness of cultural processes in the operations of the economy of the city has turned attention to the specificities of the economy and the situatedness of economic rationality within the city. Understanding the culture–economy of the city is the latest move in an analysis that reads economic activity in an expanded realm of work, leisure and taste and as socialized rather than individual. I argue, adapting Dewey, that it involves a re-evaluation of economic rationality, taking on a broader idea of instrumentalism as a form of social communication. These developments in the analysis of the operations and effects of economic rationality are traced using the example of the rise of the new middle class and the gentrification of the city.

In Chapter 7, 'In city hall' I look at the relationship between rationality and urban planning. The chapter considers the various schools of planning – rational comprehensive, radical, deliberative and postmodern – in relation to rationality and urban space. It suggests how deliberative planning may capture more of the situated practices of everyday life that were ignored in the modern city of comprehensive planning (and shown through Holston's (1989) work on Brasilia). I explore the deliberative turn and the use of the idea of communicative rationality as a tool for planning (especially through the work of John Forester 1989, 2000). The last part of the chapter considers the prospects of the related 'pragmatic turn' in planning, a development that links to the broader aims of the book. I aim to advance calls for a pragmatist planning by suggesting a role for planning as a form of argumentation.

In Chapter 8 'Cosmopolitan reason and the global city' I explore the links between cosmopolitanism, rationality and professionalism in the context of 'global' urban spaces. The degree to which a new middle class, in possession of professional skills and knowledge and decontextualized cultural capital, displays a transversal rationality is debated. Transversal rationality is a logic of transition between recognized irreconcilable spheres of rationality (Schrag 1992; Welsch 1998). This rationality of transition exists in the smooth 'space of flows' (Castells 1996) between global cities. This contrasts with the 'space of place' (Castells 1996) that makes up the rest of the city that is home to a form of practical reason that is limited by context and disconnected from power. In contrast to this divided city I suggest a broader view of communication and deeper idea of communicative (transactional) rationality results in an idea of cosmopolitanism that is as much about the encounter with difference in fully lived space of the 'ordinary' urban neighbourhood, as it is about mobilities between transnational situations.

The book offers a series of different communicative sites, from the body through to the space of the global city, on which dominant forms of rationality and communicative alternatives are registered. The point is that these sites do not exist in some sort of simple cognitive and power hierarchy, but are constantly in tension. The sort of sensory intelligence provided by body-mind is at work in the higher reaches of urban planning and in the workings of the global economy. Stark instrumentalism and extraordinary aesthetic experience, communicative rationality and instrumental rationality, system and lifeworld, abstract and lived space – all exist in a continuum, not as separate spheres. The space to best understand that continuum is the city. It is in the city where transition between communicative realms is most possible and where the diversity of lifeworlds (difference) might still be gathered into a transactional rationality of more meaningful experience.

2   On the body

RATIONALITY AND THE DISCIPLINED BODY

In western philosophy the body has been the inferior other of the mind. It has at various times been associated with instincts, habit, woman and nature. Elevated above the body is the mind, the locus of reason, culture and transcendent values. This dualism was realized in western city space from the nineteenth century in the constitution of the public and the private. The public was a realm beyond bodies, a male space of cognition and debate. The iconic figure of the modern city, the flaneur, moved through the city like a ghost, disembodied, observing the heterogeneous activities of the metropolis without any tactile involvement (Wilson 1991). In contrast private space was female and static – the home of nurturing bodies.

It is possible to see the mind/body dualism as the city ridding itself of bodies. In the Greek polis the body was celebrated as the epitome of human excellence and full citizenship. This was recouped for a time in the Renaissance but certain influences of Christianity meant that the body was thought of as low, fleshy, sinful. In the modern era Georg Simmel saw how the inhabitants of late nineteenth century Berlin closed down their bodies in public as a form of indifference, protection against the overstimulation of the city (Simmel 1950). They were unable to deal with each other emotionally. Le Corbusier's (1971) plans and buildings treated the city as a machine in which bodies circulated like abstract atoms. The pulse and press of the city is captured by Henri Lefebvre's rhythmanalysis (Lefebvre 1996). The city is full of imperatives, one of the most obvious of which is the movement of large numbers of bodies on the mass transit systems. These bodies and their particularities were lost from the planning imagination: the city was decorporealized.

In contrast the city was all about head. Le Corbusier's vision was to use the plasticity afforded by modern building materials to create a city of open, endless vistas, composed of glass and light, a transparent place. Vision is the sense most closely associated with rational faculties and Le Corbusier was projecting a rational vision both on the plan and in the heads of the inhabitants of his Ville Radieuse. The city and its inhabitants would be literally enlightened.

There is of course another reading of bodies and rationality in the city, part of a wider philosophical turn against reason itself. This is provided most strongly by Michel Foucault's explorations of the institutions of modernity as rationalising and disciplinary devices. In modernity the body has been opened up for the circulation of power and the disciplinary apparatus. Careful attention has been paid to bodies. Foucault begins his now famous discussion of panopticism by describing the decree issued in a plague town to restrict the movement of bodies.

> First a strict spatial partitioning; the closing of the town and its outlying districts, a prohibition to leave the town on pain of death, the killing of all stray animals; the division of the town into distinct quarters, each governed by an intendant. Each street is placed under the authority of a syndic, who keeps it under surveillance; if he leaves the street, he will be condemned to death. On the appointed day, everyone is ordered to stay indoors: it is forbidden to leave on pain of death. The syndic himself comes to lock the door of each house from the outside; he takes the key with him and hands it over to the intendant of the quarter; the intendant keeps it until the end of the quarantine.
>
> (Foucault 1977: 195)

Foucault goes on to discuss the construction of the panopticon in the modern prison. This has the effect of subjecting prisoners to a surveillance that they cannot anticipate – the guards might be watching them at any moment. The consequence of this is that the prisoners act as though they are being watched all the time. They discipline themselves according to the logic of the overall disciplinary regime of the prison. Foucault's point is that modern social administration has had the same effect on bodies in society at large. The example of the plague town is prophetic in this sense. From land use zoning, through to the proliferation of CCTV cameras, the modern city operates more and more like the panoptica, a carceral city (Davis 1990). In the city the body is thoroughly rationalized. It is submerged in calculation of rational efficiency (from utilitarianism), maximum return – in this case the control of bodies – for minimum investment in the technology of surveillance.

Yet even for those who acknowledge the dominance of rationalization, the body is still seen as a potential site of resistance to rationality (and this includes Foucault himself). The idea of resistance is put most powerfully by Henri Lefebvre.

> The architectural and urbanistic space of modernity tends . . . towards . . . homogeneous state of affairs . . . Everything is alike. Localization – and lateralization – is no more. Signifier and signified, marks and markers, are added after the fact – as decoration . . . it is also the space of blank sheets of paper, drawing boards, plans, sections, elevations, scale models, geometrical projections and the like . . . A narrow and

desiccated rationality of this kind overlooks the core and foundation of space, the total body, brain, gestures, and so forth. It forgets that space does not consist in the projection of an intellectual representation, does not arise from the visible-readable realm, but that it is first of all heard (listened to) and enacted (through physical gestures and movements).

(Lefebvre 1991: 200)

Lefebvre is attacking the mentalistic space of Cartesianism, based on an abstract two-dimensional geometry of points and lines. Against these spaces of representation Lefebvre sets the representational space, lived space.

Representational space is alive: it speaks. It has an effective kernel or centre: Ego, bed, bedroom, dwelling, house: or: square, church, graveyard. It embraces the loci of passion, of action and of lived situations, and thus immediately implies time. Consequently it may be qualified in various ways: it may be directional, situational, or relational, because it is essentially qualitative, fluid and dynamic.

(Lefebvre 1991: 42)

For Lefebvre the privileged site of representational space is the body. Spatial practices are lived directly before they are conceptualized. Space is produced through mimesis. Using the simile of the spider, the spider spins a web, building space outwards via tactile orientation involving symmetries and asymmetries. This space does not cohere to the consistencies and expectations of abstract, rational space. It is labyrinthine and subversive. The body resists the rationality of the city while at the same time being dominated by that rationality.

BODIES WITH CITIES

There are other conceptions of the relationship between the body and the city that consider bodies as neither rationalized nor as some sort of authentic space of escape from rationalization. This intermediate position sees bodies and urban space as mutually constitutive. Thus Grosz (1995) argues that the city is one of the crucial factors in the social production of sexed corporeality.

For Grosz the body is 'a concrete, material, animate organisation of flesh, organs, nerves, muscles, and skeletal structure which are given a unity, cohesiveness and organisation only through their psychical and social inscription as surface and raw materials of an integrated and cohesive totality' (Grosz 1995: 104). There are two prominent interpretations of the relationship between bodies and cities. The first is external and causal where the city is seen as the reflection of the naturalized body. Humans make cities but bodies are mere tools in the making, the mind is the repository of the plans and designs of the city. This is problematic for Grosz because it subjects bodies to minds and conceives of the relation between bodies and cities as one-way. The second relation is one of parallelism in which body and city are seen as analogous, especially in terms of the body politic. This emphasizes the rule of reason and the functional integration of all the parts through a disciplinary logic.

The relation between bodies and cities, Grosz suggests, is neither causal nor representational, but combines elements from each. Bodies and cities co-produce each other. Neither are distinct entities but they define one another in emerging assemblages, or groupings of activity. Grosz points to the electronic revolution as radically restructuring the interrelationship of cities and bodies. It involves potentially a collapsing of space into time as computers take on human qualities and bodies become more cyborg-like. The sorts of associations that Grosz is describing are endlessly emergent and productive.

I argue in this chapter that another perspective on bodies and cities is possible. It sees the relationship between bodies and cities as neither representational nor causal nor endlessly emergent, but rather transactional. Transactional body relations are organised in certain ways and involve relationships between the rationality of the social body and the speculative, performative aspects of the rationality of historical agents. This approach is represented most strongly in the work of the pragmatist John Dewey.

BODY-MIND AND THE CITY

Dewey too tries to overcome the mind-body separation that I am arguing sets the body against rationality or the body as a respite from the city. Dewey's idea of body-mind is constituted out of precisely those relationships that are so symptomatic of contemporary urbanism – namely the meshing of technology and bodies, diverse forms of communication, and a more fragmented experience of time-space.

> . . . body-mind simply designates what actually takes place when a living body is implicated in situations of discourse, communication and participation. In the hyphenated phrase body-mind, 'body' designates the continued and conserved, the registered and cumulative operation of factors continuous with the rest of nature, inanimate as well as animate; while 'mind' designates the characters and consequences which are differential, indicative of features which emerge when 'body' is engaged in a wider, more complex and interdependent situation.
>
> (Dewey 1958: 285)

This is an emergent theory of mind (Dewey 1958: 271). It is a mind that emerges in interaction that is at once physical, psycho-physical and mental. The physical level is one of need satisfaction for existence. The psycho-physical is a set of organized responses: 'physical activity has acquired additional properties, those of ability to procure a peculiar kind of interactive support of needs from surrounding media' (Dewey 1958: 255). This separates animate life forms from the inanimate.

> A sessile organism requires no premonitions of what is to occur, nor cumulative embodiments of what has occurred. An organism with locomotion is as vitally connected with the remote as well as with the nearby; when locomotor organs are accompanied by distance receptors, response to the distant in space becomes increasingly prepotent and equivalent in effect to response to the future in time. A response toward what is distant is in effect an expectation or prediction of a later contact.
> (Dewey 1958: 256–7)

This relationship to space is precognitive, but complex. Its complexity is at its height, I suggest, in the time-space compressions and distantiations of the modern metropolis. Space and sensation, and from it emotion, are mutually constitutive. As Dewey argues:

> complex and active animals *have*, therefore, feelings which vary abundantly in quality, corresponding to distinctive directions and phases – initiating, mediating, fulfilling or frustrating – of activities, bound up in distinctive connections with environmental affairs. They *have* them, but they do not know that they have them. Activity is psycho-physical, but not 'mental,' that is, not aware of meanings. As life is a character of events in peculiar condition of organisation, and 'feeling' is a quality of life-forms marked by complexly mobile and discriminating responses, so 'mind' is an added property assumed by a feeling creature, when it reaches that organized interaction with other living creatures which is language, communication [emphasis in original]. . . . The distinction between physical, psycho-physical, and mental is thus one of levels of increasing complexity and intimacy of interaction among natural events.
> (Dewey 1958: 258, 261)

This view of the organization of organisms has important implications for the ordering of activities and our understanding of communication and the constitutive relation of both to space. Nearness and farness are a result not of Cartesian distances between atomized bodies, but are a constitutive element of the nature or 'quality' of the transactions between organisms. This is the heart of Dewey's idea of body-mind and it has myriad implications for the way we conceive the body operating, especially in the time-space conditions of the modern metropolis. First it acknowledges the importance of the mutual constitution of space and body-mind. It orients space around the body, just like Lefebvre's spider. Unlike Lefebvre it does not rely on the Nietzschian hyper-aestheticized idea of excellence and animal naturalness. Some responses of body-mind rely on animal sensitivity and feeling (as defined above) but others involve knowledge of those responses in distantiated time-space. Farness in space and time and mediation in communication are mutually constitutive of a mind in a moving body-mind that must anticipate, form expectations and inhibit certain responses in order to form a coherent response to others in different times and spaces. This attribute is rationality, a rationality embodied and embedded in biological/technical as well as social processes.

The body is at the centre of the constitution of rationality. And so is space. Rationality is the ability to connect spaces. If we take George Herbert Mead's (1934) arguments for the intersubjective constitution of self – its origins lie in the gesturing body. Communication is built out of particular contextual gestures, which through mediated interaction form into symbols (gestures that mean the same thing to the sender and receiver). Symbols are gestures that call forth the response in the receiver that the sender anticipated. This is exactly the situation that analyses of strategic rationality deal with. How does an individual, with their own preferences, act when what results is also reliant on how others act who themselves are trying to judge how others in the situation will act. Rationality is a device for coordination of activity (though not necessarily cooperation) that relies on consistency of response. It is a rationality, even in this strategic form (instrumental, analysed by game theory) that is social and emergent from interaction. Consistency in Mead's conception relies upon the individual, the 'I', putting itself in the attitude of the 'generalized other', the whole group to which they belong.

> Rationality lies in the social process in which self-conscious individuals take themselves as related parts of their social situation in responding to its role as "generalised other". In a non-problematic or ongoing situation, this generalised other prevails as its common structure of roles in symbolic forms to be assumed and observed: in a problematic or delayed situation, this role is introduced by the individual involved as the situation is reconstructed by them in terms of the methodic process, the role (of generalised other) of the society of science
>
> (Kang 1976: 48)

The body has a special place here. It is the locus of primary sociality. Simple physical gestures allow participants to establish communication. These gestures also help the individuals to establish body schema (see also Merleau Ponty 1962). This is the overall image of their bodies such that they can be seen as objects so that gestures can be seen and understood by the partner in communication. The body is also the site of impulses. These are common to both human and non-human actors. The critical distinction is that human impulses can be inhibited or delayed. Impulses are mediated processes of implementation that are ongoing and incomplete.

Particularly crucial are hands and the central nervous system. Hands are very flexible instruments of communication. In terms of the genesis of gestures they are dextrous and expressive. They are also implements of other implements – they manipulate tools – be it the grip on the spear or the fingers running over the computer keyboard. Following from this they are intermediaries in activity: they help constitute physical things.

> The hand is responsible for what I term physical things, distinguishing the physical thing from what I call the consummation of the act. If we took our food as dogs do by the very organs by which we masticate it, we should not have any ground for distinguishing the good as a physical thing from the actual consummation of the act, the consumption of the food. We should reach it and seize it with the teeth, and the very act of taking hold of it would be the act of eating it. But with the human animal the hand is interposed between the consummation and the getting of the object to the mouth. In that case we are manipulating a physical thing. Such a thing comes in between the beginning of the act and its final consummation. It is in that sense a universal.
>
> (Mead 1934: 184–5)

This leads to a discussion of the social genesis and nature of the physical thing (Mead 1938: 119–39). In many ways this anticipates more recent discussions of the sociology of things (Urry 2000) and the way that things can act back (Latour 1987; Callon 1991). According to Mead the constitution of things as things is in terms of their social capacities. The hand is also the instrument of further instruments. The physical object socially constituted represents an implement or device for other types of activity.

> The world of physical things we have about us is not simply the goal of our movement but a world which permits the consummation of the act. A dog can, of course, pick up sticks and bring them back. He can utilise his jaws for carrying, but that is the only extension possible beyond their actual utilisation for the process of devouring. The act is quickly followed through to its consummation. The human animal, however, has this implemental stage that comes between the actual consummation and the beginning of the act, and the thing appears in that phase of the act. Our environment as such is made up out of these physical things. Our conduct translates the objects to which we respond over into physical things which lie beyond our actual consummation of the immediate act.
>
> (Mead 1934: 248–9)

There are distinct parallels here with actor network theory (ANT: Latour 1987; Callon 1991). Callon (1991: 135) states that 'actors define themselves, and others, in interaction, in the intermediaries they put into circulation'. The intermediaries can be other human beings or non-human actors, such as animals or texts, machines and money. This expresses the radical symmetry of ANT in that objects and animals, as well as humans, can act and have affects. Mead considers intermediaries as implements in the consummation of the act. But these physical things, socially constructed, have the capacity to exceed the particular situation and this leaves them with the capacity to have effects that are unanticipated. Mead and ANT share an expanded view of interaction involving significant intermediaries.

TRANSACTIONAL BODYING

In all this the body is the locus of intelligence, both cognitive and non-cognitive, habit as prior will, and emotions as the momentum of intelligence and enquiry. Thus Dewey argues:

> The conclusion is not that the emotional, passionate phase of action can be or should be eliminated in behalf of bloodless reason. More 'passions' not fewer, is the answer. To check the influence of hate there must be sympathy, while to rationalize sympathy there are needed emotions of curiosity, caution, respect for the freedom of others – dispositions which evoke objects which balance those called up by sympathy, and prevent its degradation into maudlin sentiment and meddling interference. Rationality . . . is not a force to evoke against impulse and habit. It is the attainment of a working harmony among diverse desires. Reason as a noun signifies the happy cooperation of a multitude of dispositions, such as sympathy, curiosity, exploration, experimentation, frankness, pursuit – to follow things through – circumspection, to look about at the context etc. etc
>
> (Dewey 1922: 196)

Dewey couches his idea of rationality in the melioration of diverse impulses, habits and intelligence in a range of emotions. This is the equivalent of the fullest possible exchange of the qualities of everyday life. Their fullest possible expression and melioration comes in circumstances of a robust democracy: the unhampered enquiries of a community of scientists. Rather than truth emerging from the substantive grounding of validity claims in language that aims at consensus, it comes from uninhibited speculative enquiry involving the broadest idea of communication.

The conserving aspects of organic activity that are invested in the body do not limit that energy to the individual body. Dewey argues that organisms live 'as much in processes across and "through" skins as in processes "within" skins' (Dewey and Bentley 1991: 119). This is a process of transaction, used to 'indicate the dynamic, co-constitutive relationship of organisms and their environments' (Sullivan 2001: 1). The idea of transaction is important in distinction from interaction. Interaction for Dewey implies that nature of relations between two fully formed and distinct organisms. But as Sullivan, drawing on Dewey argues: 'organisms, such as humans, are not "located" within the epidermis in an isolated, self-contained way: they are instead constituted as much by things "outside" the skin as "within" it, as well as by the skin, or site of transaction itself' (Sullivan 2001: 13).

There are profound implications of Dewey's idea of transactional bodies.

> The process of organisms living in processes that reach across and through skins illustrates the dynamic quality of transaction without suggesting a formlessness in which entities are in such flux that they have no stability, order, or identity . . . In its rejection of static ontology, the notion of transaction does not reify flux or complete dissolution of identity. Put another way, the concept of transaction no more supports a process metaphysics than it does a substance metaphysics.
>
> (Sullivan 2001: 13)

Instead, as Dewey claims, transaction involves something like 'a stability that is not stagnation but is rhythmic and developing' (Dewey 1958: 25). Dewey treats the body as an activity, a bodying, that is not reducible to its physicality.

Body is in transaction with mind and with the world. An organism's bodying does not occur in random ways but is patterned. It is patterned by the influence of habit which is styles of activity that organize its impulses (Sullivan 2001: 30). Habit is not merely repetition but something far more active. It is 'an acquired predisposition to *ways* or modes of response' (Dewey 1922: 32; emphasis in original). These patterned behaviours include thought as well as physical activity. Much of thought is not conscious but it supports conscious thought. Dewey also distinguishes between habits that are intelligent and ones that are routine. Habits 'are not within the individual but formed in and through the organism's transactions with its various environments' (Sullivan 2001: 33–4). Sullivan goes on to draw an analogy between Dewey's notion of habits and Foucault's idea of discipline; both emphasize 'the ways in which various cultural, institutional, and other nonindividual and nonpersonal structures actually constitute organisms' (Sullivan 2001: 35). She adds that Dewey's idea of habit supplements Foucault's work by looking at the 'forms of discipline at the level of organisms effected by them' (Sullivan 2001: 35). The understanding of power as habit also allows consideration of 'how the incorporation of discipline as habit can be a positive tool of the transformation of modes of discipline' (Sullivan 2001: 35).

The disciplinary environment relates to the dispersion of dominating discourses. Judith Butler's work (Butler 1993, 1997) suggests how bodies are discursive through and through in the sense that language has materiality. Butler argues that to posit the idea of a non-discursive body merely discursively constructs sex as somehow naturalized and therefore not open to political intervention. The placing of the body outside discourse is a discursive strategy that supports the naturalization of sexual difference to support gender inequality. Here, as Sullivan argues, Butler is rejecting the spectator theory of knowledge that in this case treats the sexed body as something to be discovered. This resonates with Dewey's own rejection of philosophy as the mirror of nature. Butler argues that 'doing gender' involves constant referral and repetition but also requires everyday performances to reproduce the norm. Any slight slippages or infelicities (in speech act terms) between performer and audience can result in misappropriations and misunderstandings. These slippages can result in a slightly different course of action/discourse being taken. However minute this change it may be compounded through time to result in gender being done slightly differently. It is here that transformation can come about.

In contrast to the performative view of bodies and their discursive construction, transaction offers another understanding. At the moment of performativity communication is one way, whereas transaction implies a mutual constitution. Sullivan also argues that Butler neglects the 'had' body at the expense of the 'known' body and fails to investigate the effects of discursivity in concrete situations.

> The positing of a nondiscursive body is itself a discursive phenomenon, but the discursivity of bodies does not have to preclude concrete acknowledgements and analyses of bodily transactions, including the non-reflectively lived experience of bodies. Embracing bodily discursivity does not have to entail the sacrifice of attention to the experience of bodies, and appealing to lived experience is not necessarily to attempt to get 'behind' culture to a nondiscursive ground. Thus when I urge that bodies be understood as transactional bodying, I urge both that the discursivity of bodies be acknowledged and that room be made for discussion of the concrete 'had' experiences of lived bodies.
>
> (Sullivan 2001: 61)

Butler's account is, Sullivan argues, restricted to an epistemological approach to bodies. Sullivan calls for an investigation into the effect of discourses on bodies, including bodies non-consciously 'had' as well as known bodies.

TRANSACTIONAL BODYING AND URBAN 'SITUATIONS'

Transactional bodying is deeply engrained in the creation of situations. These are 'events of mutual participation of particular potentialities for shaping space and time. The "qualities" we recognize in situated things within those situations are not in the organism or in the environment, but "always were qualities of interactions"' (Langsdorf 2002: 151). Coming from Dewey's idea of aesthetic practice, the aim at:

> expanding comprehension . . . calls upon a constitutive rather than representative understanding of communicative experience's task, in

order to direct participants towards the unsayable: to what cannot be transmitted because there is no antecedent to transmit. It requires calling upon the enduring presence of nondiscursive, somatic experience to expand out discursive parameters, and so relies upon Dewey's thesis that somatic experience – which is 'had' but not 'known' . . . continues its efficacy throughout all modes of communication.

(Langsdorf 2002: 152)

Dewey's idea of aesthetic practice focuses on the creative process rather than the artistic product of the work. The product of art and the work of art are not the same thing. Again the relationship between the two is not causal or representational but transactional: 'the actual work of art is what the product does in and with experience' (Dewey 1987, cited in Langsdorf 2002: 153) and aesthetic 'designates an orientation within experience that appreciates that doing' (Langsdorf 2002: 153). This replaces the predominant static with a processual view of aesthetics. Aesthetic experience is consummatory and connective. It unites diverse experiences and processes. But it unites in a situational rather than a transcendent way. Its situational and processual qualities mean that it encompasses difference in a number of ways. 'The diversity of this plurality ([of] matters, body; meanings, mind) assures that the unity of a work of art is not a uniformity: it is a dramatically harmonious blending of differences, rather than a reduction or purification to one stuff' (Langsdorf 2002: 153). It is constantly in process: 'experience is a matter of the interaction of the artistic product with the self. It is not therefore twice alike for different persons . . . It changes with the same person at different times as he brings something different to a work' (Dewey 1987, cited in Langsdorf 2002: 153). 'Even the "consummatory closure that marks a work of art is but a temporary stasis"' (Langsdorf 2002: 153). 'Every movement of experience in completing itself recurs to its beginning . . . but the occurrence is with a difference . . . Every closure is an awakening, and every awakening settles something. This state of affairs defines the organisation of energy' (Dewey 1987, cited in Langsdorf 2002: 153).

What the contemporary pragmatist philosophers of feminism and communication (Langsdorf, Rosenthal, Siegfried, Sullivan) are suggesting is that an expanded view of communication, encompassing the somatic and the aesthetic realms within a body-mind continuum, is communicative rationality in its fullest sense. It is a rationality that is not just about talking and thinking (the emphasis that Habermas gives – see chapter 5). It involves the body (or rather bodying), gesture and other forms of communication that are alert to messages that are not easily represented in speech and text. These messages are part of the hum and buzz and deeper organizational rhythms of the city. What I argue is that the range of communications in transaction in the city is a resource for rationality, rather than being a constantly emergent hum of activity that has no particular direction. Messages go astray, or are misappropriated, or actions are performed out of routine (by the historical 'I' against the conventional 'me') and these fuel the more speculative side of rationality. The time-space conditions of the contemporary city mean that the situations in which organisms transact are not just accounted for by co-presence, the focus of the first Chicago School, but rather the degrees of transactional distantiation or compression.

Transactional compression or distantciation is the productivity of power. I have argued this elsewhere in terms of network power (Bridge 1997b). Distantiation or distance stimulation can enhance rational capacities by inhibiting impulses but when power circulates broadly and thinly (by capillary action as Foucault put it) holding in responses becomes self-discipline and rationalization. On the other hand, over-compression of space into subjectivity gives too strong an image of control and direction. Compressed subjective space is dismissed by Butler and others and the myth of interiority. It can also be seen, I would argue, in the projective phenomenology of Merleau Ponty (1962), in which the neutral body is the transcendent communicative orientation of everyday life. And of course there is the rational choice actor, with clear purposes that she tries to pursue in a strategic context that also shares the myth of over-subjectification. It is also present in Habermas's version of communicative action. Whilst his idea marks a postphilosophy of consciousness he still relies on strongly cognitive processes being able to come to an agreement over the (stable) meaning of words.

I suggest that Dewey's transactional logic of body-mind sees the power-subjectification relation as one of tension between distantiated and compressed space. The 'vague sense of thick agency' (Rosenthal 2002) is vague in the sense that it comes from the circuitousness of diverse transactions (what White (1992) in network terms has called ambage) and vague in the sense that it involves the rhythms of the transactional bodying and psychophysical organization as well as mind. It is thick in terms of being a situational complex of diverse impulses. It is also the tension between the 'I' and the 'me', the expectations of the generalized other and the unpredictable historical agent. Where these impulses have some kind of transactional organizational pattern, that is rationality in its fullest and deepest sense.

Bodying functions in a number of ways. It provides the source of gestural communication out of which other forms of communication are built, but also a series of somatic transactions which offer communicative opportunities, particularly in experimentation and speculation, that are strong sources of new ways of communication and new meanings. The 'had' body can be a reservoir of new possible meanings when verbal communication fails, when it is impossible to put things into words. Here is the possibility for novelty, for a new direction, for transformation. These situated communicative opportunities we might see as the free spaces, or, to adapt Foucault (1986), the heterotopias that give the possibilities for resistance (we shall consider the urban spaces for tactics understood by de Certeau in the next chapter).

THE 'HAD' BODY AND SOCIAL DIVISIONS

What, though, if the resources of the 'had' body, whilst being pre-discursive, are still not free of social distinction or division? The work of Pierre Bourdieu is important in this respect. Bourdieu (1977, 1984, 1991) potentially offers an understanding of the body that conveys strongly the divisive and embodied aspects of power, especially class power. This is reproduced through the habitus. Habitus is an array of inherited dispositions that condition bodily movement, tastes and judgements, according to class position (Bourdieu 1984). The habitus is individually embodied and a shared body of dispositions – a form of collective history (Jenkins 1992). It has a good deal in common with Dewey's idea of habit, although for Dewey habit is a more active constitution of will that can be changed in significant ways that are discussed later in this chapter. The shared dispositions of habitus are instilled in the body and are non-conscious. The individual body is able to improvise in uncertain situations but the range of possible improvizations is still determined by the habitus. In the same way class power is instilled in the body beyond its ability to perform speech acts that make a difference. Body shape and gait, the movement of the limbs and the style of delivery, inton-ation and stress of vocal communication, merely highlight the distinctions of class power rather than being the basis for some communicative consensus (à la Habermas) or performative slippage that may alter the doing of class or gender (in the way suggested by Butler).

So much of the city seems to reflect Bourdieu's habitus. If the marks of class are inscribed on the body, then the city is where those signals are used most readily. In a city full of strangers, the look and movement and situation of the body in space is the social signal of status par excellence. As Simmel (1950) and later Sennett (1996) have so eloquently argued, in a setting where all you know is what you see, what you see becomes discriminated to a fine degree. The experience of the urban life has been the closing down of the body. The body is the bearer of inherited dispositions and even improvis-ations of action or speech are not able to overcome class habitus.

Bourdieu's practice-oriented understanding of human affairs also has much in common with Mead's work (for an examination of the links see Aboulafia 1999). The conserving role of habitus is like Mead's 'me', the agent acting in anticipation of the response of the entire community. That anticipation is the operation of a form of social rationality as we have seen. In that sense habitus is, if you like, the realm of rationality, or practical reason that Bourdieu observes.

TRANSACTIONAL BODYING AND TRANSGRESSION

Yet Bourdieu seems to under-estimate the communicative and transform-ative possibilities of the body. Butler herself accuses Bourdieu of a form of social determinism (J. Butler 1993, 1997). She argues that Bourdieu ignores the performative aspects of the reproduction of norms. Body performance gives off an excess that cannot be fully contained by the norm to which it is responding. Butler suggests that there is transformative potential in the fact that not all habits are fully bedded-in, and that sedimented performances can be transformed by being enacted in different contexts to the ones in which they were formed (Sullivan 2001: 98–9). As Butler puts it 'the efforts of performative discourse exceed and confound the authorising contexts from which they emerge' (J. Butler 1997: 159). The excess of the speaking body to interpellation 'remains uncontained by any of its acts of speech' (Butler 1997: 155). Sullivan argues that Butler does not elucidate on the notion of performative excess, which, in the way that is it deployed in Butler's theory, takes on transcendent qualities. From a pragmatist perspective Sullivan argues that the excess should be understood in terms of concrete situations and practices. A more concrete idea of body excess comes from Dewey's notion of impulse, that raw body energy that has the ability to disrupt established habits. Impulse does not have an existence separate from habit however: habit is the organization of impulses. Performative excess cannot be taken as something that exists apart from habit. Butler's second response to sedimentation Sullivan finds more promising. This is the possibility that sedimented performances might be transformed by being enacted in contexts different from the ones in which they were formed.

Social performativity in non-consonant environments leads to an idea of the body and the city that is more transactional in nature. As Sullivan expresses it: 'An account of performativity that recognizes the social and contextual nature of its logic can be found in the notion of habit as constitutive of transaction' (Sullivan 2001: 104). This transaction is not with a singular supporting environment however. The multiplicity of environments in which single bodies as organisms transact is a feature of modern life and especially of city life. The transactions occur across organic boundaries and via non-human agents and technical systems (Callon 1991: Latour 1987). The distantiated, emergent nature of these assemblages would suggest that human subjectivity is dispersed or distributed in these networks (Ansell Pearson 1999). This notion of a weak subjectivity seems to be compounded by an urban environment where much of what takes place is automatic, from routine software applications, to automatic doors, to intelligent traffic systems. These imperatives cohere at different levels of ordering of our thoughts and emotions. The ordering of environmental imperatives is made clear by the experience of travellers arriving in a strange city.

> Just arrived in the old colonial city, we watch the cobblestone streets and the centuries-old houses parade before our eyes, a spectacle suspended in the tropical sunlight. We stop before a low stone building with blind façade and august studded door. A plaque informs us that this was the Palace of the Conquistador, built in 1502. We see that it is now a hotel. We step inside the timber-ceilinged entrance hall, are led up the majestic

staircase to the upper floor, and admire, like a set in some historical movie, the great oak table under the chandelier.

Then we take a room. The spectacle of the lofty room and its colonial furnishings now thickens with the smell of old wood and damp stone and the fabrics of the bed cover and draperies, and with the tangible density of their substances. The spectacle becomes real. From its window the street extends to the main square, and to the left, to the cathedral. Established now in this haven of reality, we can begin to envision the layout of the city, of the country, and of the country beyond it we had left to come here.

(Lingis 1998: 41)

The directives of the environment means that much of what bodies do is the response to the levels of directives, rather than the workings of consciousness and cognition. The sense of subjectivity here is found in the responses to the orderings of the environment as a result of the motility of the body and its relationship to other bodies through Merleau Ponty's idea of projective intentionality via the anonymous body. When I see another body in action, the objects around it take on an importance distinct from my own assessment of them. Those objects 'are no longer simply what I myself could make of them, they are what this other pattern of behaviour is about to make of them' (Merleau Ponty 1962: 353), the meaning of which is accessible to me because of bodily existence.

As Sullivan argues, the problem with projective intentionality via a neutral body as a way of understanding the other is that it takes no account of the already inscribed differences in bodying that this form of projection may mis-recognise and in fact obstruct understanding. This is in part what Bourdieu is getting at in his view of social divisions that are below discourse (Bourdieu 1991). Sullivan is pursuing Dewey's idea of transactional rather than projectional phenomenology of the body. Bodies and objects are transactionally constituted. This is closer to Butler's dismissal of the idea that there is some kind of body that stands outside discourse or culture. Unlike Butler's idea in which bodies come with excess because they are bodies (a transcendent view of excess), transactional bodies create novelty in concrete situations (situational excess). That is also, as I take it, not to say that the transactions of bodying are radically open to reconstitution of objects and bodies. Habit has a strongly conserving momentum that in many cases is the embodied effects of power that Foucault analysed so well. But neither does it suggest that the body is wholly responding to imperatives in the environment in a way that underestimates the degree to which bodying is co-constitutive of environment.

The reason that Sullivan likes Butler's second promise of performativity of displaced speech acts is that it looks much more to the environment–organism relations, rather than relying on individual body excess, that she (Sullivan, and Dewey on which she draws) argues is much more subsumed as impulses under the organizing hand of habit. Dewey in fact has an idea of habit as an exercise of will.

> The essence of habit is an acquired predisposition to ways or modes of response, not to particular acts except as, under special conditions, these express a way of behaving. Habit means special sensitiveness or accessibility to certain classes of stimuli, standing predictions and aversions, rather than bare recurrence of specific acts. It means will.
>
> (Dewey 1922: 42)

Dewey's idea of habit cautions against Deleuze and Guattari's (1988) ideas that see bodies, machines and other bits becoming effects in rhizomatic surfaces. Rhizomatic surfaces are like plant root systems that interconnect horizontally but have no vertical hierarchy. Habit as power shapes unfolding transactions in certain directions and according to established patterns. But neither do habits constantly reproduce social divisions in ways that, Bourdieu suggests, even encompass improvisations. What Dewey looks to is the overall relationship of organisms to their environment. Humans (as well as animals) operate in increasingly complex environments, of which the city is the pre-eminent example. The range of environments makes for a range of trans-actions of organism and environment via different habits. As Sullivan explains it:

> Because adults transact with many different sorts of social and cultural institutions, their habits constitute a variety of different and potentially conflicting dispositions. In isolation, each of an adult's particular habits may be inflexible because of the rigidity of the institution that helped form it, but precisely this rigidity can help create the impetus for change when it causes a particular habit to conflict with others. Taken individually, an adult's rigid habits bode ill for her potential to change the gender structures in society. Taken as a whole, however, her habits make up a complex web of overlapping habits in which individual habits begin to wear upon and challenge and influence each other. When they do so, the resulting friction between and weakening of some habits disrupts the usual ways adults habitually transact with the world, opening up possibilities for reconfigurations of habit and thus of culture as well. And eventually, with an increased number and variety of habits that are interrelated, one's habit-forming itself may become subject, as Dewey says, 'to the habit of recognising that new modes of association will exact a new use of it'. Thus habit is formed in view of possible future changes and does not harden so readily.
>
> (Sullivan 2001: 104–5).

The expanding and diverse interconnection of habits starts to instil a habit of adaptability. The melioration of diverse habits is the operation of rationality.

Habit supplies some of the energy of rationality but there are other forces in play which are to do with here-and-now performance, affect as well as broader experimental-technological activities. Rationality is the mediation and melioration of all these forces. It becomes more critical the more forces there are and the more distanciated and mediated their connections. We have already seen through Mead the significance of delays in communication. This applies not just to traditional ideas of rational communication in language but non-verbal and emotional communication too.

SPECULATIVE PERFORMATIVITY IN THE CITY

Interactional vandalism and social division

The disturbance of ongoing habits can result from the impact of speech acts that are out of place and present a problematic situation for the receiver of the communication. This effect is made all the more powerful by non-discursive elements of transactional bodying that disturb ongoing transactional routines.

The significance of routines of spoken communication and the impact of different speech acts is the stock-in-trade of symbolic interaction, ethno-methodology (Garfinkel 1967) and the detailed ethnographic work of the early and later Chicago Schools. An exemplary example of this work is Duneier and Molotch's (1999) study of the communicative interaction between homeless African-American men and passers-by on certain side-walks of Greenwich Village in Manhattan. But before considering Duneier and Molotch's analysis of what they call the 'interactional vandalism' wrought by the homeless men on the (mainly) white middle class women passing by, it is worth considering the transactional bodying already in place before the encounters unfold.

Problematic transactional bodying is most evident where bodies are in some senses out of place. In the city this is represented most strongly by the communicative affect of the homeless body. The homeless body subverts normative sense of the public and the private that has been constitutive of the meaning of the city. Homeless people do things with their bodies in the street that have traditionally been confined to the private realm. As Wright (1997: 58) argues: 'people living on the street are not just neutral bodies, but subjugated bodies and resisting bodies moving through, sitting, lying down, and sleeping in, the social-physical spaces of the city, a negative trope for surrounding housed society.' This is especially acute for homeless women. As Watson puts it:

> homeless women's bodies can be seen to represent a challenge to the feminine body, the mother or wife located in the home, cooking in the kitchen, going about her daily tasks. In a sense she comes to be the feared 'other', held up as a counterpoint to a happy 'normal' life. As

such the homeless woman serves to keep housed women in their place, by her presence she becomes a reminder to all women of what they might become if they step out of line. By sleeping in the street wrapped in a blanket, by bringing her bed into the street as it were, she is also starkly disrupting the public/private boundary on which much planning regulation is based. What we see in a graphic way is the private, and the sphere associated with feminine domesticity and sexuality seeping in to the public in disruptive and threatening ways.

(Watson 1999: 96–97)

Grosz (1995), Gatens (1992) and Watson (1999) have argued in different ways that the body is constituted at the intersection of the subject and the social. The homeless body serves a broader rationality of discipline of the social body by being posed as a threat and exemplar of life outside the norm. The threatening nature of homeless bodies on the street is part of the normalizing discourse that is constitutive of home, and masculine and feminine bodies.

Within this wider rationalization bodies are also constituted by trans-action and situation. The transactional constitution of the passers-by in Duneier and Molotch's (1999) study comes from the intersection of a range of situations. There is the need for polished appearance and accomplished and purposeful deportment that accompanies the transaction of the middle class body. For the middle class women control is compromised by the male gaze (a patriarchal relation). Furthermore these are female bodies in public and subject to the fear of male violence that is constitutive of certain city spaces. This heightened guardedness is compounded by the transactions between black and white bodies so poignantly portrayed by Anderson (1990) in his exploration of the street interaction between poor blacks and white gentrifiers, discussed in Chapter 3.

Duneier and Molotch's study of African-American homeless men's con-versational interactions with female, white middle class pedestrians on Sixth Avenue in Greenwich Village, Manhattan, reveals a range of rationalities at work. First there are the rational expectations that come with interaction between strangers in the street (discussed more fully in the next chapter). Conversational routines, particularly the opening and closing of convers-ations, are 'technically' infringed by both the men and the women on the sidewalk. The homeless men break the norms of civil inattention by certain conversational devices. In response the women infringe normal codes of politeness in response to conversational requests in order to close down the interaction. The infringements are acts of 'interactional vandalism' but as Duneier and Molotch argue, they are not random, irrational acts of violence. Nor are they acts of violence that are conventionally policed. Nevertheless they act on a certain rationality which is all about the men getting them-selves recognized, challenging the established street routines that mirror the deeper social divisions when it comes to the transaction between the homeless body and the wider society. In terms of the analysis throughout this book they are seeking to establish the 'I' against the social norm of the 'me', the generalized other of societal expectation, in order to get a sense (however minute) of power. This is a rational oppositional move. These speech acts out of place are in this sense potentially transformational but the conversational and wider threat to the existing order that they pose prevents any broadening of horizons and in fact prompt rejection and closure. Conversationally the basis of trust is undermined (even though that trust supports inequalities) in a way that is particularly complex for the women. First there are the sexist ways greetings and parodies of intimacy – 'I love you baby', 'Hey pretty' – and the ensuing conversational techniques that deny full control of the interaction and play on the vulnerability of the female body in public. This combines with middle class attitudes and the particular situation of Greenwich Village, noted for its liberal and bohemian attitudes. It was of course the site of Jane Jacobs' (1961) famous celebration of the diversity of urbanism.

The complexity of the interaction is revealed by subsequent conversations that Duneier and Molotch had with certain pedestrians. As Laura expressed it: 'I think what you said about our white liberal guilt is true . . . I was thinking about the fear in my voice and I think you are right about people having a hard time saying no, that you're not supposed to say no.' As a result of interactions like this Laura felt that she was 'getting more conservative as I get older and it's not taxes. . . . I guess I don't mind some of the street cleaning up around here . . . [and I] HATE my reaction' (emphasis hers: Duneier and Molotch 1999: 1287).

Duneier and Molotch's example is one of the layered divisions of 'race', gender and class. In this case the potentially transformative elements of bodies and speech out of place does result in a recognition (that Laura agonizes over) but the expectations of the generalized other that are socially divided result in a transaction of transgression and threat that closes down the communication and reinforces the ongoing habit of social division. Indeed, as Duneier and Molotch point out, it is these reactions built from street situations that in part explain wider support of zero tolerance and the more brutalized reactions to the homeless in New York and other cities. The situation of speculative rationality is part constitutive of whether there is transformation or reinforcement of the status quo. The speculative aspect of transactional rationality can go either way.

Situationist derives

The transformative possibilities of performatives 'out of place' in the city were explored as a deliberate political strategy by the situationists. The situationists looked for the dissolution of the distinction between art and everyday life (Plant 1992). They engaged in a political strategy of purposeful disorientation to reveal the ideological strictures of everyday life that had become subordinated to the demands of commodity capitalism. These strategies had a strongly urban flavour. In the early 1950s the Lettrist International (including Guy Debord, Gil Wolman and Michel Bernstein), a group that were subsequently to influence the establishment of the Situationist International, produced a periodical *Potlatch*. They wanted to transcend the distinction between revolutionary politics and cultural criticism. They focused their attention on the context of everyday life. They called for a unitary urbanism, 'a critical study of the city utilising all artistic and technical resources' (Plant 1992: 56–7). The situationists argued that the situations in which people live in the city are created for them. They are numbing of experience. The situationists wished to become psychogeographers to understand the precise ways in which urban space shaped the behaviour and emotions of individuals (Plant 1992: 58). One key device was the 'derive'. The derive involved movement without specific goals in which 'one or more persons during a certain period drop their usual motives for movement and action, their relations, their work and leisure activities, and let themselves be drawn by the attractions of the terrain and the encounters they find there' (Debord 1957: 50, cited in Plant 1992: 58–9). As Plant argues, this was playful and speculative. It was a conscious inversion of the way that the urban environment instilled certain 'states of mind, inclinations and desires, and to seek out reasons for movement other than those for which an environment was designed' (Plant 1992: 59).

This was not a rejection but an embracing of technology to enhance everyday life rather than repress it. 'We have invented the architecture and the urbanism that cannot be realised without the revolution of everyday life – without the appropriation of conditioning by everyone, its endless enrichment, its fulfilment' (Kotanyi and Veneigem 1961, cited in Plant 1992: 61). Plant describes

> Chtcheglov's remarkable 'Formulary for a New City' [which] experimented with 'a thousand ways of modifying life' . . . [he] considered the possibilities of the mobile house, changeable city environments, and the establishment of such areas as the 'Bizarre Quarter', a 'Happy Quarter', a 'Sinister Quarter', and advocated the 'changing of landscapes from one hour to the next' which again would result in 'complete disorientation'.
> (Chtcheglov 1958: 3–4, cited in Plant 1992: 61)

Flash mobbing and glamouring

Activities that use urban space to create surprising performances to unsettle the taken for granted routines of daily life have continued in various guises, the most recent of which is the contemporary phenomenon of flash mobbing. Although closer to the non-reasoned derives of the surrealists than the more deliberately rational and conscious strategies of the situationists, flash mobs do have the ability to unsettle established urban rhythms. These inexplicable gatherings of people broke out in cities across the USA and Europe in 2003. UK newspaper The *Guardian's* New York correspondent received an email instructing him to place himself at 7pm in one of four designated bars (dependent on month of birth) and buy a drink and act casual. Thereafter a mob representative would turn up with other instructions. At 7.28 the mob should disperse. No one should remain at the mob site after 7.30. Younge and 100 other people born in January, February or March, found themselves in Puck Fair bar. Staff are wondering why they are suddenly rushed off their feet. People have been contacted by web magazines and via email. Younge takes up the story:

> Shortly after 7 a young woman turns up with further instructions on small slips of paper. We have to leave Puck Fair by 7.16 and go up the road to the fancy shoe shop, Otto Tootsi Piohound, by 7.18 and disperse by 7.23. As we make our way out of Puck Fair a swarm of several hundred others, born later in the year, flood out of various bars and head for Otto Tootsi. Our arrival at the shoe shop creates instant confusion. A few moments ago there were less than 10 people there; now there are a few hundred. The assistants look utterly confused. 'What do you think is going on' I ask one. 'I don't know' he shrugs. 'Maybe we have just become very popular all of a sudden.' One of the managers is less accepting of the intrusion. He rushes to the door to try and close it but when the mob keeps coming he gives up. Meanwhile the shop fills with mobsters acting out the role-play on their slips. 'You are: On a bus tour from Maryland,' it says 'You are excited but also bewildered. It is as if the shoes were made in outer space. If you have a cellphone, dial a friend. Say 'Guess where I am?' After a pause, say, 'In a SoHo shoe store.' Or: 'In one of those New York City mobs'.
>
> (Younge 2003)

Another flash mob gathered in the carpet department of Macey's in Manhattan all claiming to be co-residents of a warehouse outside Manhattan who only made purchases jointly. They spent 10 minutes in negotiation with a bemused salesperson when they asked for a 'love rug' to play on.

In Denver on 11 August 2003:

> At 5–43, the mob of about 250 – including people from a wide range of ages and backgrounds – was split into two groups, one on the upper floor of the Pavillions, the other on the ground. The mob, in unison and with remarkable precision, suddenly began counting down from 60. Then both sides started a virtual ping-pong game that lasted for nine minutes, with one side yelling 'ping' and the other returning serve with 'pong'.
>
> (www.smartmobs.com/archives/001453.html) [13 August 2003]

There are a range of different types of mobs advertised on the smartmob website administered by Howard Rhinegold, author of *Smart Mobs: the Next Social Revolution* (2002) which advocates mobbing being used to bring people together. One such is glamouring, organized by a wiki. Glamouring is a pun combining the French for love with the prefix gl for glitter. People meet in a publicly accessible area, and then break into small groups to spread messages of love ('Honour thy children', 'Cherish!'), hope ('Love prevails!'), peace ('Hug thy neighbor!') and personal responsibility ('Strive dangerously!', 'Think!') and fun (messages like 'Don't walk . . . tango!' on a street corner).

> While waiting for groups to form or for a friend to join you, participants can donate supplies and ideas to people to 'make glamours' . . . people who can't join the touring groups are welcome to join in the art portion of the project. . . . Bring inexpensive beads, string and charms to make permanent fits for people met along the way or to leave hanging just within sight as a gift for people who rarely look around and at their world. (I usually tag mine with a label that says 'Take me!' so it's not left alone because of thinking someone lost it!). Bring markers and post-its or index cards to make hand-out messages and 'Free Hug' cards and similar. Bring washable sidewalk chalk, cake glitter (environmentally friendly biodegradable edible glitter) and anything else that will make a low impact, high message statement while sticking to the legal codes of peaceful assembly and 'not' vandalism. Dress for the weather, wear a costume if you like, bring snacks of drinks if you need them, and bring good walking shoes.
>
> (www.kinhost.org/wiki/pmwiki.php/News/
> GlamouringForWhirledPeas03) [13 August 2003]

The spirit of playful creation that was at the heart of the situationist derive was of course tied to a broader social critique of the conditions of everyday life in commodity capitalism. Smart mobs involve a mix of people with a range of motives for being there. What they reveal is the power of speculative actions and the presence of a group of bodies to disrupt routine urban rhythms. These events are reminiscent of ethnomethodological strategies to disrupt the routines of everyday encounters in order to reveal something of their structure (Garfinkel 1967). They also signal the transformative possibilities of bodies in motion in unusual ways, bodies engaging in actions that are out of place. They have the potential to disturb settled habits (of both mobsters and their audience or those with whom they interact). They invoke discordance that broadens the experiential field and provides a wider repertoire of habits. They play on the speculative 'I' of rationality in the broad range of environments that the city affords.

Raves and clubbing

The power of novelty to introduce new habits that require reconciliation with discordant or conflicting habits is also seen in the case of youth movements. Dewey (1938) lauded youth as a group whose habits had not settled and showed greater flexibility. There was greater possibility of experimentation with cultural styles and bodily presentation. The urban dance movement of the 1990s is a good example of experimentation and eclecticism over the music, the event and the bodily styles and displays. The rave scene showed the effectiveness of rapidly organized gatherings and the social experimentation that came out of these happenings. Raves had no fixed location but were organized via phone, email and word of mouth.

The urban dance scene came out of clubs, such as Cream in Liverpool and the Hacienda in Manchester, that rapidly gained an international reputation. Playing a mixture of techno and house, the music itself, often computer generated, sampled and mixed hitherto distinct music styles, such as heavy metal and soul into a novel pastiche. Bennett argues that rather than being, at least in its beginning, a commercially manufactured style foisted on the youth market, the eclecticism of the music matched the record collections or musical memories of the clubbers, who relished the mix (Bennett 1999). Bennett uses this and other evidence to suggest that the sub-cultural theory of youth music movements was misplaced. The sub-cultural theory of youth movements was developed by the Birmingham school of cultural studies to account for music styles and body fashioning as a resistance to the prevailing structural socio-economic conditions that these young people faced. Thus, the punk movement, with its associated fashion made from everyday items (bin bags, safety pins to hold patched-up clothes), was a working class reaction to the divisions of a consumerist society, by inverting worthless objects into body statements of identity. There is undoubtedly a good deal of truth in this. The lyrics and actions of bands such as 'The Clash' and 'The Sex Pistols' were deliberately anti-establishment and sought to represent the plight of unemployed working class youth.

The sub-cultural theory that the Birmingham school was applying was a development of the Chicago School work on a sub-cultural theory of urbanism. The size and diversity of the city permitted the growth of distinct sub-cultures that displayed different cultural norms (sometimes labelled deviant, or bohemian). These different norms were able to develop because the city was fragmented into what Park famously called 'a mosaic of little worlds' (Park 1969). The worlds were neighbourhoods with distinct ambiances, lifestyles and norms. In Deweyian terms there were distinct complexes of organisms and their environment.

Bennett's point about urban dance music is that the association between a distinct environment and sub-cultural response cannot be drawn so clearly. Drawing on Maffesoli's (1996) work, Bennett argues that the urban dance scene was more 'tribal' than sub-cultural. The alliances were less permanent, less bounded and more complex and overlapping than the sub-cultural thesis allowed for.

As Bennett indicates, the greater complexity and fluidity of these sub-cultures had already been questioned by McRobbie (1994) in her critique of the sub-cultural analysis for its focus on public activities (on the street, in the club, at the gig) and on men. Looking at the youth scene for teenage girls revealed a situation of greater complexity. The home, and especially the bedroom, was an important site for the discussion of music and fashion with friends over teen magazines with an overlapping sense of identities and allegiances. For the older teenagers the urban dance scene provided an environment distinct from the typical heterosexist gaze of the traditional disco pick-up scene. The mass atmosphere and non-intrusiveness of the dance floor of raves and urban dance, allowed the women to share a vague sense of dispersed eroticism. This was potentially empowering and trans-formative in gender terms.

The previous examples suggest that the city is full of spaces in which body-mind transactions are potentially transformative as well as iterative of the social status quo. Slippages in speech acts, bodies out of place, the situational excess of bodies and the possibilities of thick agency can all alter the way social norms of gender, race and class are done. Yet the city has also been seen as the environment that is pre-eminent in suppressing the transformative potentials of collections of body-minds. The self-limitation that comes with over-subjectification is apparently no more obvious than in the interaction between strangers on a city street. This has been confirmed by the Duneier and Molotch example. Indeed Simmel over a century ago saw the city street as the location where a new form of urban rationality ordered the relations between bodies and minds (Simmel 1950). It is to the city street that we now turn our attention.

3   On the street

In the previous chapter we considered the rhythms of bodying as part of the broadest conception of communicative rationality in the city. In this chapter we begin by limiting ourselves to the narrowest idea of rationality. It is the instrumental, highly cognitive and strategic version that is assumed to dominate in urban society. This form of urban rationality is most evident in relations between strangers on a city street. I suggest that even in this restricted situation strategic rationality is leaky, and suggestive of wider and deeper communicative repertoires. Even where action is instrumental and strategic, that is not necessarily concerned with values, the need for strategic advantage tends to make participants open to communicative signals in excess of that directed by the situation of the encounter itself. Even in this limited case there is a tendency towards communicative expansion. Communicative expansion can push against the established norms of interaction, the conformity to which Mead called the 'generalized other' or the 'me' of interaction. This is the rationality of expectations governed by the generalized other of the community.

In contrast there are the effects of the historical agent that also expands communication in different ways. This is the manifestation of the 'I' rather than the 'me' and it represents what I call the more speculative side of rationality. Speculative rationality is often evident at the edges of interaction and enunciation. Walking and encounters on the street can also be at the edges of communication. This is shown by Anderson's (1990) famous study of street relations in poor a African-American neighbourhood, next to a gentrified white district in an Eastern American city and which I discuss in detail. De Certeau's (1984) work is an intriguing investigation of these edges of communication, which I explore in the second section of the chapter. The rest of the chapter is concerned with exploring examples of communication on the street and the creation of new meanings in the city. I compare the examples of communication in the transgressive space of Mardi Gras in New Orleans and Sydney. But whereas Sydney Mardi Gras is confronting and transformative (Bruce *et al.* 1997), in New Orleans Mardi Gras transgressive acts (such as disrobing in public) merely reinforce prevailing gender norms (Shrum and Kilburn 1996; Jankowiak and White

1999). The effect of performance art in transforming street relations is explored through Rossiter and Gibson's (2000) analysis of the Urban Design Capsule in Melbourne. Reversing communicative norms is also shown by Jones's (2000) study of the street protest of Jackie Smith, who occupies the street as a private space in public protest at the use of the civil rights museum on the site of the Lorraine Motel where Martin Luther King was assassinated in 1968.

To begin with, I explore the communicative realm that has traditionally been seen by urbanists as the one where rationality is at its height and communication most restricted: the encounters between strangers on a city street.

THE MASK OF RATIONALITY

According to Simmel (1950) the multiple stimuli of the city would overload the typical urban dweller to which the psychological response is a levelling-out of the types of responses to stimuli, including the stimulus of other human beings. Although 'metropolitan life . . . underlies a heightened awareness and a predominance of intelligence in metropolitan man' (Simmel 1950: 410), the range of stimulations meant that intellectuality was accompanied by a growing incapacity, or lack of energy, to differentiate between urban experiences, a pervasive *in*difference. This flattening out of experience goes hand in hand with the psychological power of money to reduce hitherto qualitative distinctions between phenomena to quantitative ones. Intellectuality and the calculative aspects of a money economy go hand-in-hand: 'Money economy and the dominance of the intellect are intrinsically connected. They share a matter-of-fact attitude in dealing with men and things' (Simmel 1950: 411). Relations between people become less instinctive and emotional and more calculative and objectified. As Simmel (1950: 411) puts it: 'All intimate emotional relations between persons are founded in their individuality, whereas in rational relations man is reckoned with like a number, like an element which is itself indifferent'. Everyday social action becomes increasingly rational and less emotional, 'the conceivable elements of action become objectively and subjectively calculable rational relationships and in so doing progressively eliminate emotional reactions and decisions which only attach themselves to the turning points of life, to the final purposes' (Simmel 1990: 431).

In conditions of multiple stimulation in the metropolis the individual 'self' is preserved by engaging in rational, rather than emotional, relations with others. Expressions of self are limited to the types of information that could be transacted like money. As Sennett (2000: 567) expresses it: '. . . this Simmelian notion of the mask of rationality is that information is exchanged . . . but communication is lowered, particularly communication which transcends self-interest as well as communication of an emotional sort.'

Urban rationality is therefore a form of self-repression. Its principles of operation seem to be very close to instrumental rationality as conventionally understood. It confines the information exchanged to representations of self-interest, to instrumental purposes. Deeper value positions remain hidden. It concerns 'neutral exchanges between people in a state of equilibrium amongst strangers' (Sennett 2000: 566). Those strangers are merely treated as conditions or means to self-interested ends. The terms of engagement are often characterized by mutual suspicion and competition.

STRATEGIC RATIONALITY AND THE SPACES OF THE CITY

Strategically rational expectations most surely underpin the routines of indifference that urbanists from Simmel onwards have said is symptomatic of the modern city. Mutually rational expectations of indifference result in repertoires involving the avoidance of any eye contact beyond a navigational glance, maintenance of body distance and avoidance even in micro- spaces and overall body gloss (Goffman 1971). These paralinguistic orientations in public are well established in western cities.

Self-limitation in micro-spaces is paralleled by the settings of different types of interaction in the city as a whole. At the street-scale interaction cues are given about the rules of interaction that can be expected. This relates to the overall socio-spatial segregation that characterizes many western cities. Expectations as to the types of participants in the interaction are given by the location in which that interaction is occurring. Certain types of people 'belong' or are expected in particular parts of the city. Space connotes identity in the modern city just as clothing and appearance marked out differences in the socially more diverse settings of the pre-modern city (see especially Sennett 1990; Lofland 1985, 1998). Social homogeneity in the separate neighbourhoods of the city means that all too often routines of interaction are locationally and socially specific. As Sennett argued some time ago (1970), homogenous communities (in terms of class and ethnicity) can have comfortable regimes of interaction amongst themselves whilst harbouring a childlike psychology about the characteristics of people in other parts of the city that encourages stereotyping at best and demonization at worst. That is why Sennett is so right to look to those borderland spaces of the city where people routinely encounter others that are unlike themselves. It is in these circumstances that Simmel suggested routines of reserve and indifference would be established: the masks of rationality would cover the faces of the urban dwellers. I suggest that an environment of borderland spaces and social heterogeneity are indeed where such routines operate but they are also where these routines are most likely to 'give off' other tacit messages. Borderlands are also where such routines can face moments of undecidability or feelings of strangeness. It is in these sorts of spaces that the strategic possibilities of encounter are at their greatest, not their most limited.

WALKING THE CITY: FIRST STEPS IN STRATEGIC RATIONALITY

The multiple forms of psychic stimulation that take place on a walk down a busy city street, the numbers of strangers that must be communicated with in order to walk without hindrance, relies on a structure of communication that enables people to move on without infringing the privacy of others (Henderson 1975; Karp 1973; Collett and Marsh 1974). To negotiate the strangers on the sidewalk people need to be able to signal to each other and interpret the messages they are sending by eye contact and body movement, in a consistent way (L. Lofland 1985; J. Lofland 1976; Wolfinger 1995). Actions are interdependent. At its most trivial level whether I choose to move to the left or the right to avoid bumping into you will depend on whether I believe you are going to move to the left or the right. Furthermore I assume that you are trying to avoid bumping into me and so are making the same sorts of calculations, choosing which way to go based on how you think I will move. It's not even as simple as that. You must make your decision based on the assumption that I am also trying to judge how you will move and knowing that I am trying to choose knowing that you are also trying to anticipate my choice and vice versa. The fact that these layerings of anticipations get resolved successfully in a split second, thousands of times a day for the average city dweller shows that they have become tacit, almost intuitive. The point is that even though there are routines of interaction to rely on, strategic rationality is still active within these routines – they are tacit but not necessarily habitual. To maintain a strategic awareness strangers need to be sensitive to the additional information that is 'given off' even in the most tacit interaction. What at first might seem like 'rather neutral exchanges between people' – a glance in the street between strangers for instance – are much richer than the limited nature of their operation might seem to convey. They are forms of tacit coordination that can have considerable communicative import. That is evident when the structure of communication is played upon in some way, however trivially. Playing with the structure of communication can have its own communicative content. Here is the case of tacit but asymmetrical communication given by the game theorist Schelling (1980).

> . . . a motorist at a busy intersection . . . knows that a policeman is directing traffic. If the motorist sees, and evidently sees, the policeman's directions and ignores them, he is insubordinate; and the policeman has both an incentive and an obligation to give the man a ticket. If the motorist avoids looking at the policeman, and cannot see the directions, and ignores the directions that he does not see, taking a right of way he does not deserve, he may be considered only stupid by the policeman, who has little incentive and no obligation to give the man a ticket. Alternatively, if it is evident that the driver knew what the instructions were and disobeyed them, it is to the policemen's advantage not to have

seen the driver, otherwise he is obliged, for the reputation of the corps, to abandon his pressing business and hail the driver down and give him a ticket.

(Schelling 1980: 149)

In this context of the 'legal status' of communication Schelling points to Goffman's research on 'face work', citing Goffman (1955):

> Tact in regard to face-work often relies for its operation on a tacit agreement to do business through the language of hint – the innuendo, ambiguities, well placed pauses, carefully worded jokes and so on. The rule regarding this unofficial kind of communication is that the sender ought not to act as if he had officially conveyed the message he has hinted at, while the recipients have the right and the obligation to act as if they have not officially received the message contained in the hint. Hinted communication, then, is deniable communication.
>
> (Schelling 1980: 224)

This layering of communication is a result of the mutual anticipations based on strategic rationality. As Goffman describes it:

> Decisions made by directly orienting oneself to the other parties and giving weight to their situation as they would seem to see it, including their giving weight to one's own. The special possibilities that result from this mutually assessed mutual assessment, as these affect the fate of the parties, provide reason and grounds for employing the special perspective of strategic interaction.
>
> (Goffman 1970: 101)

The communication in these kinds of interactions is tacit but informative. One of Schelling's significant insights is that the type of rational structure that guides tacit communication has an important influence on more explicit forms of communication (Schelling 1980: 102–11). For example the tacit moves shown by the example of the motorist and policeman is paralleled by explicit communication and performance that plays on the structure of communication. Goffman uses an example (picked up by Schelling) of street stemmers (salesmen) 'who commit themselves to a line that will be discredited unless the reluctant customer buys, the customer may be trapped by considerateness and buy in order to save the face of the salesman and prevent what would ordinarily result in a scene' (Shelling 1980: 128fn). I discussed a more recent example in the previous chapter given by the research of Duneier and Molotch (1999). They showed how a number of black homeless men on the streets of Greenwich Village, from a position of relative powerlessness, continuously usurped the rhythms of expectation in the structure of conversation between themselves and the people passing them on the street. These challenges to interactional expectation (what Duneier and Molotch call 'interactional vandalism') are particularly complex when the disempowered position of being homeless and black is crosscut by interactional games that play on the gender talk that reveals more ambiguous power status of middle class women pedestrians.

The examples discussed above show how strategic action can be used to deceive but the very basis of deception is the use of strategic action to save the face of the other. According to Goffman (1967) drawing on Durkheim (1976) such minor courtesies are abbreviated versions of social rituals. Even in the non-intimate social setting of the modern city such small moves are made to preserve the sacredness of the other person. As Goffman (1967) argues, 'face work' serves to maintain a line in a flow of events. Maintaining a line gives a certain credibility to actions. It also helps express the character, or at least an impression of the character of the agent. The more distanti-ated relations of the modern city mean that social spheres are fragmented and can stretch over different ranges of time and space. The fragmented nature of the urban personality performing roles in different sphere of activities was one of the early concerns of urban theory (Simmel 1950; Wirth 1938).

Fragmentation of self can happen within an interaction as well as between spheres of interaction. Rather than being a source of anomie, this fragment-ation of self can be an explicit tactic in interaction as Goffman observed:

> . . . the 'splitting' character of the self during interaction, that is, the general capacity of an individual to handle himself by stepping back from what he seems to have become in order to take up adjustment involving distance from this person; and that, in turn, this capacity results from the inevitable interactional fact that that which comments on what has happened cannot be what has happened. I here attempt to derive a property of interactants from interaction. My claim is that the individual is constituted so that he can split himself in two, the better to allow one part to join the other members of an encounter in any attitude whatsoever to his other part.
>
> (Goffman 1971: 117)

The splitting of the self in different realms of interaction is, in some senses, a form of hybridity – the self as a composite of networks of association (Bridge 1997b). The range of social spheres can be seen as arenas of the pursuit of instrumental interests that require the adoption of various 'lines' in interaction and the maintenance of 'face' in the spheres of interaction. These lines are to some extent artificial. However the losing, saving and keeping of face is a matter of considerable emotional and intellectual import. They are also emotionally important in the preservation of the face of the other. To see another lose face or be discredited has emotional impacts on all around. Actors will often move to preserve the face of the other even if that person is the offender. The discomfort felt by the middle class women passing the homeless men (Duneier and Molotch 1999) was in part the impossibility of rescuing the interaction in terms of prevailing norms of ritual and remedial work. The combined effect of the rule of self-respect and the rule of considerateness is that the person tends to conduct herself during an encounter so as to maintain both her own face and the face of other participants. This means that the line taken by each participant is usually allowed to prevail, and each participant is allowed to carry off the role he appears to have chosen for himself. A state where everyone accepts everyone else's line is established (Goffman 1967: 11).

The acceptance of 'lines' and credibility without trust can help expand the public realm of communication. From a different perspective Sennett (1974) points to the importance of rhetorical performances in certain spaces of the city. The adoption of the communicative mores of the theatre allowed dis-cussants to transcend social differences by using an 'as if' form of communic-ation. Theatrical modes of address gave rhetorical force to statements but also helped to cross social class by stepping out of normal relations and performing as actors – deliberately following artificial lines (discussed below).

If we take the adoption of lines and roles as Goffman suggested them, then all forms of initial social interaction are on an 'as if' basis. In the modern city spheres of interaction involving actors taking various parts or lines is constitutive of everyday interaction. What is significant is that this 'as if' deployment of character in interaction is mutually recognized and assumed in the city. Actors work to preserve the consistency of the line adopted by the other actor (and to correct any faults) knowing that the other actor is performing similarly. The mutually assumed role-playing is underpinned by strategic rationality. It involves the mutual acknowledgement of the 'line characters' adopted by the actors. This is what Goffman calls 'ritual' – a regard for others and the roles they perform.

THE SPACE FOR SMALL STEPS . . . AND BIG LEAPS

Some of the most important conventions that keep interaction in equil-ibrium are spatial. Goffman distinguishes between spaces of the self and spaces of encounter – situations, occasions, gatherings, singles, 'withs'. In the spaces of the self, comprising body space, stalls, use space, sheaths, possessional territory (see Goffman 1971), the individual seeks refuge and respect from others. These spaces reflect the information that an individual can expect to give off and protect about self in public. This is the informational preserve, 'the set of facts about himself to which an individual expects to control access while in the presence of others' (Goffman 1971: 39). There are also conventions over the protection of conversation between groups in the spatial realm. This conversational preserve is 'the right of an individual to exert some control over who can summon him to talk and when he can be summoned; and the right of a set of individuals once engaged in talk to have their circle protected from entrance and overlooking by others' (Goffman 1971: 40).

Space is significant at the edges of communication, in the ways that communication gets established. These conventions in space require the sequencing of understanding the strategic purposes of others. This includes the 'critical sign', 'the act on the part of the other that finally allows the individual to discover what it is the other proposes to do' (Goffman 1971: 13). The critical sign is followed by the 'the establishment point', 'the moment both parties can feel that critical signs have been exchanged regarding compatible directions and timing, and that both appreciate that they both appreciate that this has occurred' (Goffman 1971: 13). This mutual acknowledgement relies on strategic rationality. It is a way of establishing a provisional security that allows further interaction or the removal of attention elsewhere. Its baseline conditions are shown by pedestrian traffic. As Goffman argues:

> City streets, even in times that defame them, provide a setting where mutual trust is routinely displayed between strangers. Voluntary coordination of action is achieved in which each of two parties has a conception of how matters ought to be handled between them, the two conceptions agree, each party believes this agreement exists, and each appreciates that this knowledge about the agreement is possessed by the other. In brief, the structural pre-requisites for rule by conventions are found.
>
> (Goffman 1971: 17)

These conventions are established by the conditional mutual agreement, the small steps established by strategically rational interaction. There is also the space for big leaps that take interaction and communication in a different direction. Big leaps can come about from the intervention of historical agency, the 'I' of rationality, or through a move to a new norm as a result of social rationality (of the 'me'). Much of what I discuss throughout this book concerns historical agency. There are also big leaps of social coordination, the 'me' of rationality, which can be seen in Schelling's (1980) idea of 'focal points'.

Street rationality is sensitive to the range of discursive and non-discursive cues that can give extra information to the interacting parties. But there are times when rationality will not help coordination, even at the basic level of the coordination of pedestrian interaction. Where there is no rational coordination possible with the available options participants might seize on elements in the situation to help them coordinate with each other. This element introduced from outside the interaction Schelling called a 'focal point'. Finding a 'focal point' on which rational expectations can converge helps establish a new equilibrium to allow further communication to proceed. It is a strategic move to alter the context of interaction and choice. In these situations of public encounter strategic rationality is at its height. Focal points are usually highly situational. They are qualitative aspects of the inter-actional environment that have conspicuousness at the time of communic-ative crisis. The city is full of surprises and juxtapositions that offer new meanings and focal points.

The creative potentials of language can also lead to the possibility of unpredictable or novel statements that involve 'neologisms' on which dis-cussants might focus their attention (Johnson 1993). Neologisms are state-ments that occur out-of-equilibrium and therefore can have a focal quality and the ability to bind in participants in new ways. Novel utterances and new paths through the city can have similar focal point effects in shifting rational expectations, as we shall now see from the discussion of de Certeau's pedestrian rhetoric.

RATIONALITY AND TRANSGRESSION: WALKING WITH DE CERTEAU

> The act of walking is to the urban system what the speech act is to language or the statements uttered.
>
> (de Certeau 1984: 97)

In the above quotation de Certeau is suggesting the creative and innovative possibilities of walking the city. Speech acts use diverse elements of language, figures of speech and neologisms to make new meanings that have impacts just like physical actions. Speech acts often play on the proper use of language and in a similar way walking can play on the proper meaning of the city. A walk can thread together diverse locations and situations and otherwise disrupt the proper geography of the city dictated by the rational plan. De Certeau sees in the act of walking the possibility to experience the city differently, in the same way that 'speech acts' can change inherited meanings in language. Pedestrian wanderings in the city can help give a feeling of personal freedom that can be experienced secretly to avoid the gaze of the urban authorities and the urban order: 'The user of the city picks out certain fragments of the statement in order to actualize them in secret' (de Certeau 1984: 98). The effect here is presumably a sense of personal liberation, being the author of one's own urban story, precious because unshared. The figure that most emphatically combines stylistic innovation and a secretive existence in the city is that of the flaneur. This male, moneyed, non-working, fashion-conscious urbanite moved freely through the streets of nineteenth century Paris (and other European capitals) observing the lives of others and diverse urban situations in a passive but acquisitive way, adding the myriad of urban stories to his own in a sense of growing cosmopolitan experience. De Certeau acknowledges that not all urban dwellers can move as freely as the male and moneyed flaneur but suggests, tantalizingly, that all urban dwellers have the potential to impress their own pedestrian story on certain parts of the city. Urban practices such as walking can be a form of individual window-shopping on the urban experience.

As we have seen, Simmel thought urban social relationships were like monetary exchange in their degree of disinterestedness. These deadening and alienating conditions of interaction leave urban dwellers prone to the wider rationalities of the market and bureaucracy. For de Certeau the rationalities of the city work according to a 'scopic regime' like a panopticon, in the way that Foucault had described it (Foucault 1977). He argues that the scopic regime is most clearly manifest in the very concept of the city as a totality. The 'concept city' is characterized by the production of its own space (repressing other spatial practices) – a synchronic system of power (a 'nowhen') that overrides local traditions – and the city itself is established as a universal anonymous political entity.

Instrumental rationality is hegemonic in the concept city. The city as totality is reflected in the rational plan, a view from above the city that seeks control by treating it as a machine or a human body. The heteronymous practices of city dwellers are ignored in this scheme. City dwellers themselves are treated as objects in the means–ends rationality of the master plan and utilitarian bureaucracy – they are treated as numbers in the professional discourses of social administration. The city itself is objectified as a series of rational, functional spaces that enable the economic system as a whole to function smoothly.

Yet for de Certeau there are certain urban practices that can act as forms of resistance against this dominating logic. The most prominent is the act of walking the city. 'The walking of passers-by offers a series of turns (tours) and detours that can be compared to "turns of phrase" or "stylistic figures". There is a rhetoric of walking. The art of 'turning' phrases finds an equivalent in an art of composing a path (tourner un parcours)' (de Certeau 1984: 100). This pedestrian rhetoric works on and against 'the geometrical space of urbanists and architects' which 'seems to have the status of "proper meaning"' (de Certeau 1984: 100). So just as new turns of phrase can add novelty to language, new paths and styles of walking the city can work against the rationalist logic of the planners. It can create a labyrinth of connotations and associations that are the creative outcomes of urban practices, rather than sleepwalking the modernist master plan. Certain pedestrian experiences can come to represent larger stories in the city more widely (just as in language synecdoche is the naming of the part of an object that comes to represent the whole object – sails for ships, for example). The urban pedestrian can also take shortcuts, or splice together routes and experiences that the master logic of the concept city normally keeps apart – just as in language asyndeton suppresses conjunctions and adverbs. This can result in striking statements and juxtapositions. A slight change, such as the omission of a conjunction, can radically alter the sense of what is said (compare 'stop start' with 'stop and start', for example).

Emancipation and enunciation

De Certeau realizes the essentially personalized and cognitive emancipations that may result from walking the city – what he calls a 'discreteness' of the experience of cities: the city walker

> creates a discreteness, whether by making choices among the signifiers of the spatial 'language' or by displacing them through the use he makes of them. He condemns certain places to inertia or disappearance and composes with others spatial 'turns of phrase' that are 'rare', 'accidental' or 'illegitimate'.
>
> (de Certeau 1984: 98–9)

The discreteness of experience that de Certeau celebrates resonates with Dewey and Mead's idea of the 'I', the historical agent that 'bursts upon the scene' (Shalin 1992, 271). It stands as distinct from and in tension with the 'me' or the generalized other. The 'I' does not conform to the expectations of the symbol sharing community. What de Certeau is tracing is the assertion of the 'I' against the master rationality of the city. This tension between the 'I' and the 'me' is for de Certeau the contrast between 'style' and 'use' in action and language.

These are individualized and stylistic renderings of the city that are reminiscent of flaneurie. If we are to follow the metaphor of the speech act then stylized gestures will not 'come off' unless they are received and acknowledged by an audience. Intriguingly, de Certeau expresses the linguistic relationship between speaker and audience as the equivalent of a form of bodily location in the city:

> In the framework of enunciation, the walker constitutes, in relation to his position, both a near and a far, a here and a there. To the fact that the adverbs here and there are indicators of the locutionary seat in verbal communication – a coincidence that reinforces the parallelism between linguistic and pedestrian enunciation – we must add that this location (here-there) (necessarily implied in walking and indicative of a present appropriation of space by an 'I') also has the function of introducing an other in relation to this 'I' and of thus establishing a conjunctive and disjunctive articulation of places. I would stress particularly the 'phatic' aspect, by which I mean the function, isolated by Malinowski and Jackobson, of terms that initiate, maintain, or interrupt contact, such as 'hello', 'well, well' etc. Walking, which alternately follows a path and has followers, creates a mobile organicity in the environment, a series of phatic topoi. And if it is true that the phatic function, which is an effort to ensure communication, is already characteristic of the language of talking birds, just as constitutes the 'first verbal function acquired by children,' it is not surprising that it also

gambols, goes on all fours, dances, and walks about, with a light or heavy step, like a series of 'hellos' in an echoing labyrinth, anterior or parallel to informative speech.

(de Certeau 1984: 99)

What de Certeau is suggesting here is a kind of penumbra of communication. It is a rhythmic and mobile set of possibilities that can draw in other relations and situations, 'a mobile organicity', and one not tied down by the onerousness of informative speech acts. The phatic elements of communication are registered in speech (the hellos and well wells) but of course are also hyper sensitized in the movements and inclinations of the body. This echoing labyrinth of phatic communication will become even more significant when we consider the constitution of the public realm in Chapter 5.

De Certeau's metaphor of speech acts and pedestrian rhetoric thus takes us beyond the private city of the flaneur. He distinguishes between 'style' that connotes a singular (like the flaneur) and 'use', which is the social phenomenon through which a system of communication manifests itself in actual fact: it refers to a norm. 'Style' and 'use' both have to do with a way of operating (of speaking, walking etc.), but style involves a peculiar processing of the symbolic, while use refers to elements of a code. They are the equivalents of Dewey's 'I' and 'me' in terms of rationality. They intersect to form 'a style of use, a way of being and a way of operating' (de Certeau 1984: 100). The other element of speech acts is that to have illocutory force they must resonate with an audience. This means that they must act within norms or conventions. The convention must exist and be accepted and the circumstances must be appropriate to invoke the procedure, otherwise the speech act will not come off – it will be a 'misfire' (Austin 1962). Speech acts only come off within conventions of acceptance and interpretation.

So far the elements of pedestrian rhetoric have been used to suggest improvisations that work around the prevailing rationalist logic or 'proper meaning' of the city. Improvisations occur most readily at the edges of communication, in the use of language to establish dialogue, or what de Certeau calls the phatic function 'which is an effort to ensure communication' (1984: 99). As we explored earlier in the chapter, the routines at the edges of conversation were most thoroughly researched by the social/symbolic interactionist tradition in sociology. As Goffman (1959, 1967) and others have argued, there are certain conversational routines, forms of body presentation and interaction rituals that operate to establish, maintain and terminate communication. Goffman (1970) suggested that these fringes of interaction, often involving strangers, are governed by rational strategy and by dramaturgical expression. This is a shared imaginary realm but one that is crucial in establishing the reality of everyday life. As well as conversational routines the margins of interaction also involve body language, a form of tacit communication.

At the margins of communication involving phatic utterances and body language, strategic rationality is doubly implicated: in governing the interaction routines and the first-impression assessment of strangers entering the interaction. Yet it is also at these margins where innovation in speech (through speech acts) is most likely. Therefore to succeed, it is reasonable to assume that any innovatory speech acts will have to conform to a set of rational expectations, or a norm (see Bridge 1997a, 2000). There is good reason to believe that the norms that exist in new situations of interaction are in fact coordinations of rational expectations (Sugden 1989; Lewis 1969; Ullman-Margalit 1977). Some of them might have been attained via the big leap of focal points (Schelling 1980). As we have seen, novel words or novel elements in the environment can take the norms in a different direction (Johnson 1993), but they are still coordinations of rational expectations. New situations of interaction need rational coordination in the absence of pre-existing traditions or customs.

What all this suggests for those speech acts that are at the fringes of communication to succeed is that they must conform to a 'use', or norm (longstanding, or novel), which is built out of a set of mutually recognized (however tacitly) rational expectations between speaker and audience. 'Style' must indeed combine with 'use' to make the innovatory 'act' be appreciated, just as intention must conform to convention to make the speech act come off. Otherwise the act or speech act will, as Austin says, misfire. In the same way the walker in the city has the possibility to innovate but must do so in a way that meets the rational expectations of each urban situation she encounters.

At this point in the argument we are caught by what might be called Simmel's revenge: that in the modern metropolis the prevailing convention of interaction and communication is one of indifference, involving minimized and neutral exchanges of information among strangers. It is unlikely that innovative signals in communication could ever come off in this prevailing atmosphere. Mutual suspicion and self-repression behind the mask of rationality make the modern city very unpromising ground either for the giving off of innovative signals or their reception. People will not be open to such suggestions.

BEING 'STREETWISE'

The way that rational expectations of interaction on the street can reinforce social divisions and mutual suspicion is demonstrated amply by Elijah Anderson's (1990) study *Streetwise*. Anderson looks at street relations in an African-American low-income neighbourhood adjacent to a white gentrified district in 'Eastern City' in the USA. Anderson makes a distinction between street etiquette and street wisdom. Street etiquette involves generalized rules of conduct based on superficial categorizations of others. This is the kind of street interaction, or urban rationality that Simmel was referring to, and that

Goffman studied in such detail. Street wisdom is a more sophisticated approach gained by paying closer attention to neighbourhood persons and activities. Street wisdom is acquired in a series of stages. There is mental note-taking of people in the street and their activities at different points in the day. Repeated interaction and note-taking leads to knowing about the other person in a more discriminating typology than etiquette provides. From this observance territorial communions with or without talk can develop. Street etiquette is a rudimentary form of 'tunnel vision' when dealing with others '. . . most people come to realize that street etiquette is only a guide for assessing behaviour in public. It is still necessary to develop some strategy for using the etiquette based on one's understanding of the situation' (Anderson 1990: 230). The kind of background observation that goes into street wisdom is like field research, Anderson argues:

> Once the basic rules of street etiquette are mastered and internalized, people can use their observations and experiences to gain insight. In effect, they engage in 'field research'. In achieving the wisdom that every public trial is unique, they become aware that individuals, not types, define specific events.
>
> (Anderson 1990: 230–1)

This finer grained assessment promotes confidence in negotiating the street. Being streetwise means being able to take the upper hand. It is a very serious social game. Streetwise persons have encountered every other type of stranger and have a certain edge to their demeanor.

Anderson's study illustrates a number of things about rationality, communication and difference. The first is the constant tendency towards communicative expansion even from the most pared down communicative regimes: it is better to be streetwise than to rely on the much blunter instrument of street etiquette. Being streetwise allows greater discernment by African-Americans of different whites and whites of different African-Americans. Whilst this enables black–white friendships in certain cases, in the overall context of the interaction between poor blacks and wealthy whites, being streetwise just helps support mutual avoidance. A slightly fuller communicative repertoire that being streetwise represents, can still sustain social division in varied and subtle ways. Anderson points to some telling instances of this in his book.

> A young black male, dressed in a way many Villagers call 'streetish' (white high-top sneakers with loose laces, tongues flopping out from under creased gabardine slacks, which drag and soak up the oily water; navy blue 'air force' parka trimmed with matted fake fur, hood up, arms dangling at the sides) is walking up ahead on the same side of the street. He turns around briefly to check who is coming up behind him. The white man keeps his eye on the treacherous [snowpacked] sidewalk, brow

furrowed, displaying a look of concern and determination. The young black man moves with a certain aplomb, walking rather slowly.

From the two men's different paces it is obvious to both that either the young black man must speed up, the older white man must slow down, or they must pass on the otherwise deserted sidewalk.

The young black man slows up every so slightly and shifts to the outside edge of the sidewalk. The white man takes the cue and drifts to the right while continuing his forward motion. Thus in five or six steps (and with no obvious lateral motion that might be constructed as avoidance), he maximizes the lateral distance between himself and the man he must pass. What a minute ago appeared to be a single-file formation, with the white man ten steps behind, has suddenly become side-by-side, and yet neither participant ever appeared to step sideways at all.

(Anderson 1990: 217–18)

In terms of street etiquette this is an example of 'good behaviour' and is most common in a certain part of the neighbourhood where there is a high concentration of professional whites and where whites and blacks often encounter each other.

Anderson describes how the figure of the black male youth is the most stigmatized in public. He conveys how this discrimination affects him in certain times and places on the street. From his own field notes:

At 3.00 Sunday morning I parked my car one street over from my home . . . When I reached the corner, after walking parallel to the stranger for a block, I waited until he had crossed the next street and had moved on ahead. Then I crossed to his side of the street; I was now about thirty yards behind him, and we were now walking away from each other at right angles. We moved further and further apart. He looked back. Our eyes met. I continued to look over my shoulder until I reached my door, unlocked it, and entered. We both continued to follow the rules of the street. We did not cross the street simultaneously, which might have caused our paths to cross a second time. We both continued to 'watch our back' until the other stranger was not longer a threat.

(Anderson 1990: 219)

Anderson reflects that skin colour was important in this interaction leading to a mutual distrust, although as being male, able bodied and young they were in some senses 'well matched'. He goes on to discuss how gender affects the interaction. In the situation where young men are threatening and black young men seen as the greatest threat, many black males will take on the anxiety of the woman in the street and take preemptive avoidance measures to relieve her anxiety. These layerings of anticipations are strategically rational and communicative but nevertheless support an interactional infrastructure based on avoidance and suspicion that reproduces broader social divisions.

The visuality of this division is revealed by eye work. As Anderson points out, whites tend not to hold the eyes of a black person. The inversion of this of course was the 'hate stare' that whites gave blacks in parts of the American south during the height of the civil rights movement. In more routine interactions visual interaction between blacks and whites tends to be restricted to a simple awareness check. This is accompanied by the scowl by whites to maintain social and physical distance from young blacks.

The examples Anderson gives are where visual interaction is at its most defensive and limited. Even this most pared down, tacit form of communication works to a structure of communication that underpins much richer forms of communication. Fleeting moments of tacit communication can play off ratified forms of communication (the proper language for speech acts, the proper meaning of the concept city) because the essential structure of communication is the same for both, that is they are both based on rational expectations. As we have seen, the structure of normal communication gives the possibility for all kinds of what Goffman calls 'unratified' communication to take place, including hints, innuendos and ambiguities, that allow the interaction to carry messages over and above what is routinely going on. These nuances permit new meanings to be conveyed but they are sufficiently close to the structure of expectations of the communication to assure participants of their validity. Unratified communication is a good example of how innovations can be made between participants that have none of the surety of established routines of interaction from inherited traditions and social norms.

There is a paradox about improvisations and innovations in public (either through speech or action as traditionally understood): the broader the public to which they are performed the less likely they are to come off. Yet the narrower the communities of interest or tradition, where innovation is more likely to be recognized, the less likely it is that these will foster innovative acts because coordination relies on forces such as habit and tradition that are most resistant to change and to the cognitive work required for innovation. Narrower communities are also likely to be the least receptive to innovation because their prevailing norms of interpretation prevent unusual gestures being interpreted as anything other than strangeness or threat.

There are examples in the city of social encounters, sometimes of the most seemingly demonstrative and uncontrolled kind, that can at the same time reproduce established social norms. Equally there are occasions in which street celebration can be truly confronting and transformative. The examples are that most exuberant use of the street, for carnival, or in this case Mardi Gras. Carnival has been understood as a ritual of inversion (Bakhtin 1984; Rapaport 1999) or reproduction of existing hierarchies (Edmonson 1956). Two examples of Mardi Gras in two different cities demonstrate how out-of-place communicative acts, that are full of excess, can result in the reproduction of social norms and divisions (New Orleans Mardi Gras), or can widen the communicative realm and shift social norms (gay Mardi Gras in Sydney).

CONFORMING TO ESTABLISHED NORMS: NEW ORLEANS MARDI GRAS

Mardi Gras, the traditional pre-Lenten carnival, has developed in various cities across the world over the last two centuries. It marks the release from everyday roles and routines in a display of exuberance before the strictures of Lent. The New Orleans Mardi Gras reaches its climax in the four days before Ash Wednesday (culminating in Fat Tuesday) and the city is witness to a parade of floats through the streets as well as a host of other festivities. The floats belong to long-established as well as newer Krewes (permanent carnival societies). Krewes were originally masked and masqueraded as aristocracy. They threw gifts from the floats to the begging crowds below. In the present-day New Orleans Mardi Gras beads of different colours and lengths are thrown to the crowds on the streets (Shrum and Kilburn 1996)

One of the more recent of the 'traditions' of Mardi Gras in New Orleans is flashing or disrobement in public. This takes a number of forms. Men and women on the crowded street expose their genitalia (buttocks or genitals for men, breasts for women) in exchange for beads. Participants believe the tradition has always been part of Mardi Gras but in fact it developed in the 1970s. It is spatially restricted to certain blocks in Bourbon Street. Bourbon Street cuts across the centre of the French Quarter and has long been an alternative space, for example the burlesque clubs in the 1950s had their doors open so that passers-by could catch a glimpse of proceedings inside. Disrobement at Mardi Gras is more like flashing than stripping and it involves non-professionals, ordinary members of the public.

Shrum and Kilburn (1996) have studied the New Orleans Mardi Gras over a number of years. The background to the development of ritual disrobement they see as the banning of the parades from the French Quarter; a dramatic increase in the types and lengths of beads thrown from floats which deflated the value of short strings of beads; and innovative exhibitionism. Gay revellers on one section of Bourbon Street in the 1970s began a practice of 'weenie wagging', briefly displaying the penis over a balcony or on the street. Another proximate influence was a gathering that included nudists in a second floor apartment in Royal Street. In 1975 there was an attempt to engage the crowd below in reciprocal nudity. Beads were used as part of the enticement and whereas gay flashing was widely observed but not copied and reciprocal exposure did not take hold, the exchange of beads for exposure spread rapidly over the next few years (Shrum and Kiburn 1996).

Shrum and Kilburn argue that the offer and exchange of beads for disrobement became an 'instant tradition' because it was legitimized by broader market relations. The deviant act of disrobement is legitimized by the acquisition of capital represented by beads. It is 'the competition for wealth and its public display' (Shrum and Kilburn 1996: 434). That explains why the prevailing trope of masking (Mardi Gras is the carnival of masks) also failed to take with disrobement. People exposing themselves are unafraid to reveal their true identity (as they might by hiding behind a mask) because their performance is legitimized by its linkage to symbolic wealth. Part of the currency of that wealth is its public display and conspicuous consumption. It is important to be identified with this wealth. Masks are too discrete. The 'doing' of wealth requires both performance and consumption. 'What seems like an expression of hedonism [disrobement] is a calculated market (i.e. moral) choice to enter the symbolic economy' (Shrum and Kilburn 1996: 447). They reject Bell's (1976) argument that as capitalism loses its work ethic only hedonism remains and that hedonism is anti-cognitive and anti-intellectual, at odds with the functional rationality of the market. The performance of ritual disrobement is part of a rational economic exchange.

Drawing on Veblen's (1899, 1961) work I go on to suggest later in Chapter 7 that hedonistic displays, for example of conspicuous consumption, have their own situational rationalities. The conspicuous display of long strings of beads around the neck is the symbolic equivalent of the displays of conspicuous consumption, such as the gentrification premium involving elite gentrifiers in London and Sydney (Bridge 2001a,b).

Ritual disrobement also works on visual cues of social interaction. In the crowded and noisy streets potential disrobers have to be singled out by groups of men and women using visual clues (in Goffmanesque fashion), usually provocative dress and the wearing of long strings of beads, most typically acquired by disrobing. So across the streets of the city, market relations prevail. Market exchanges involve choice and negotiation over price.

But there are also hierarchical power relations at work. Shrum and Kilburn argue that the throwing of beads from the raised floats to a baying public represents the 'command relations' of earlier aristocratic hierarchies. Beads are thrown down by the masked aristocrats onto the peasant throng below. Catching beads is a matter of luck or a result of being picked out at the whim of a Krewe member. Being able to gain beads by disrobement is an escape from the happenstance of feudal relations and a focus on the individual efficacy of action.

Hierarchical relations are also represented in physical space by the relationship between people on the wrought iron balconies of the French Quarter dwellings and the people on the street. Balcony revellers are in a privileged position to pick out street crowd members and throw down their beads. But there is another ritual ordering that involves the balconies, one that reflects wider gender discrimination. Only women expose themselves on the balconies and only at night. The pleadings of a group on the street can extend to a whole crowd focused on a single female on a balcony. Unlike the street exchanges, which are particularised and bead payment is made after exposure, beads are hurled onto the balcony prior to, as well as after exposure. Masses of beads are thrown up, too many to catch. Performers are also able to choose who they receive beads from on the street. This vertical and symbolic hierarchy has the air of exhortation and veneration Shrum and Kilburn argue. As they depict it:

> Male partners often solicit beads or chants from the crowd, acting the role of procurer to a collectivity. Women are subjected to the commanding repetition of 'show your tits', a rhythmic, determined, sometimes thunderous challenge. A stationary core of idolaters orients toward a particular woman, pointing with outstretched arms in synchronisation to a chant that grows progressively louder until transformed into cheers at the moment of disrobement.
>
> (Shrum and Kilburn 1996: 444)

Shrum and Kilburn argue that disrobement in this context is much more a pure performance: 'balcony displays are public performances that require homage or tokens, while street exposures reflect the routine transactions of commercial life' (Shrum and Kilburn 1996: 447). Balcony displays are fetish, veneration rituals that literally put the women on a pedestal 'for men to worship and control as sexual objects' (Shrum and Kilburn 1996: 449). Women are permitted a degree of assertiveness in street transactions (in procuring disrobement or bargaining to expose) but that is subjugated to a more powerful symbolic hierarchy that controls and objectifies women.

Ritual disrobing at Mardi Gras became an instant tradition, a new focal point in the creation of traditions, to use Schelling's approach. In the highly circumscribed time–space arena of Bourbon Street over four days of Mardi Gras, ritual disrobement has become a new convention of behaviour that in the rest of the year, and in this space, is illegitimate. But rather than this new coordination of expectations changing expectations it merely reproduces in symbolic form the unequal market and gender relations that exist in wider society. New Orleans Mardi Gras in this sense is a very ordered space. It reflects wider orderings of gender and economic relations (as well as ethnic segregation reflected in the social compositions of the different floats).

New Orleans Mardi Gras is highly ordered, even for those engaging in otherwise transgressive behaviour. For the majority of the million or so visitors who attend Mardi Gras over the four days of celebration, the standard norms of pedestrian etiquette are maintained (Jankowiak and White 1999). In their observation of the 1991 Mardi Gras Jankowiak and White found that any exuberant behaviour was largely confined to within group interaction. Comparing Mardi Gras with the Christmas parade and with a standard Saturday night, interaction between strangers at Mardi Gras was at a low level and where it did occur tended to be in terms of terse sexual callings. The exchange of beads or wearing of costumes did facilitate out-group interaction but in rather individualized ways. There was no general experience of communitas for the majority of non-participants at the festival.

The festival tends to intensify interactions within the group but not between strangers (Jankowiak and White 1999).

Mardi Gras in New Orleans is highly contained and non-transgressive. The marching bands are highly ordered and organized and the floats segregated racially. The participatory rituals reflect wider economic and gender orderings performed to a passive public who limit any exuberance to within group relations

SYDNEY MARDI GRAS: CHALLENGING ESTABLISHED NORMS

Mardi Gras in Sydney is the opposite of the carnival in New Orleans. Whereas the New Orleans festival is ordered and normative, Sydney Mardi Gras celebrates excess and unpredictability. Whereas nakedness in New Orleans is a flash for cash, nakedness in Sydney is exuberant and celebratory. The sheer situational excess tends to draw in the observers. As Bruce, Murphy and Watson portray it: 'It involves the crowds by its flamboyance and outrageousness, by its spectacle and sound, by its boldness and inventiveness' (Bruce *et al.* 1997, 62).

The first march took place in on 24 June 1978 to commemorate the 1969 police raid on the gay Stonewall Bar in Greenwich Village, Manhattan, that precipitated the Stonewall Riots, a key event in the modern gay liberation movement. In the afternoon of 24 June there was a second march in Sydney intended to be non-confrontational, a celebration of gay pride. The 1000 participants became 1500 as people were called out of the bars and onto the streets. After 20 minutes the march reached Hyde Park where the police turned off the loudspeakers of the truck and attempted to arrest the driver. The marchers moved to Kings Cross, a traditional marginal space and the red light district. There ensued a riot as police at one point removed their identification badges and surged into the crowd (Bruce *et al.* 1997).

This was a defining event and resulted in a lobbying to amend New South Wales state legislation, which gave the police a free hand in defining acceptable behaviour and making the arrests at the original event. Successful lobbying to change the legislation provided a rally point in subsequent years and despite the non-involvement of the lesbian community for a number of years, the event continued to grow. By 1989 it was again Gay and Lesbian Mardi Gras and by 1994 it attracted 600,000 spectators out of a Greater Sydney population of 3.5 million.

The street activity at Sydney Lesbian and Gay Mardi Gras is a series of 'subversive topoi' (Bruce *et al.* 1997: 66) involving 'disguise, fancy dress, travesty and drag' . . . 'it parodies its own oppression' with T shirts such as "Nobody knows I'm gay" and "I'm not gay but my boyfriend is". 'Each offers the norm whilst wittily undermining it' (Bruce *et al.* 1997: 67). The communicative intent is to invert given gender and sexual norms as a way of asserting identity. This is achieved in part through parody and playfulness:

... in the 1993 parade, Batman resumed his liaison with Robin, Spiderwoman found a female Captain America, Noddy and Big Ears drove by with 'Just Married' on the back of their familiar red car. From the Flintstones, Wilma and Betty's car bore the sign: 'Drop Dead Fred: Betty's Better in Bed', Thelma was with Louise. The statements made by these four parade entries are meant not to abuse but to disabuse ... The parade displays sexual stereotyping overturned. There is dressing up and dressing down. Bodies are transformed. They are seen as opportunities for display – either the severe (generally leather) or the exuberant (generally drag). The Dykes on Bikes roar by – wearing less in 1993 than 1992, weaving and roaring, seizing the stereotyped image of the macho-male Harley Davidson bikers, parading in a phalanx, bare-breasted and buttocked, sporting shades, leathers and tattoos. Women sprout dildos and exaggerated nipples. Another group of women strip down to become human version of the mannequins on which the clothes shops hang their wares – divested of display, apparently naked but for the high heels and bouffant hairstyle. A group of men turn against male uniformity, swathing themselves in diaphanous material to become birds of display with costumes that shimmer, that move, that blow, that sparkle, that reveal, that catch the eye, that enlarge an extend and have as their cultural referents the eastern voluptuary and the harem ... the displays of Mardi Gras combine the subjective pleasure of the exotic with the desire to shock.

(Bruce *et al.* 1997: 68)

The gay muscle men represent an ironic resistance to the macho culture that results in gay bashing, some of the perpetrators of which look on from the crowd.

Although the exuberance and excess does represent a widening of communicative repertoire on the street, there is a question mark as to how clear the message is.

The extent to which the parade is 'readable' by the majority of those who watch is debatable. At a quite literal level, messages and banners are often indecipherable. Acronyms and abbreviations mean little to most of the spectators. There is rarely a relationship between the purpose of the organisation and the image offered. (The Gay and Lesbian Teachers and Students Alliance were an exception, going for academic gowns and school uniforms, but this 'text' may have been too dull to read with interest). The danger is that the parade becomes pure spectacle leaving its overt political points to be understood by the few. Bombarded with glitter and light, the straights in the crowd may simply read what they see and confirm what they always thought they knew 'queens will be queens'.

(Bruce *et al.* 1997: 68)

It is not the direct political messages that are the most important communicative event of Mardi Gras, but its ability to disarm. It is not a communication about being for one thing or against another but simply being, and in particular being able to be in a particular space. This is a point Bruce, Murphy and Watson make:

> What is demonstrated on the streets in Mardi Gras . . . is more than sculpted bodies and pretty frocks. It is a proud and determined reminder that gays and lesbians are co-tenants of Sydney space. As the chant had it: 'We're here, and we're queer and we're not going shopping. Get used to it!'
>
> (Bruce *et al*. 1997: 72–3)

The fact that the written messages on the banners, or the way that organizations are presented is not at all clear in fact may add to a broadened communicative act that is about disrupting easy assumptions. Mardi Gras represents the speculative and playful edge of activity as well as more established gay performativity. It suggests an assertion of the 'I' that is not wholly subsumed in the rationality of the community or the 'me'. Sydney Mardi Gras is not all about order. It is partly about speculation. These speculative acts might change meaning by trying out meaning in different ways. They are the leading edge of a rationality that is the balance between the 'I' and 'me' as a community establishes itself in urban space. This presence registers is less visible ways, as another effect of Mardi Gras is to associate a section of Oxford Street as pre-eminently gay space. This has been reflected in the action of the local state and the courts in dealing with objections to the activities of certain gay commercial establishments. The presence of the gay community is given as sufficient support for the amenity value of the bars and saunas in question. Oxford Street is the main axis of Mardi Gras and that physical connection gets registered in other realms of rational adjudication.

If Mardi Gras is meant to be a period of transgression, a time out of time, allowing for the fact that New Orleans is a general festival and Sydney is focused on the gay and lesbian community, the Sydney event is much more transgressive and broadly communicative. The exuberance of bodily display and nakedness in Sydney contrasts with the more mundane commercial exchange for flashing or the veneration of balcony strippers in New Orleans. Sydney Mardi Gras is much more about communication over boundaries and the establishment of a community presence that broadens the experience of all the city's inhabitants. The street allows the blurring of boundaries just as much as it reinforces wider social orderings. The speculative nature of activities in Sydney suggests a broader communicative rationality at work in contrast to the narrower economic and gendered orderings of the New Orleans events.

DISRUPTING THE STREET

The way that the street can make differing communicative impacts is also evident when it is used for specific political or artistic purposes. We conclude this chapter by considering two further city case studies. The first is the Urban Design Capsule, an artistic event in Melbourne that was used to change the nature of street interaction (Rossiter and Gibson 2000). It happened over a relatively short period and was exploratory and experimental in nature. The second, the Street Politics of Jackie Smith, involves the use of the street as a direct political act by inverting the meaning of public and private and living on the street over a period of years as an ongoing political protest (Jones 2000).

The regular international arts festival held in Melbourne in 1996 had an unusual piece of installation art.

> In a world where privacy is vanishing, come see your future. For sixteen days of the Festival, five of Melbourne's street performers will be hermetically sealed behind the glass walls of Myers [department store] Bourke Street Windows! . . . these intrepid art-stronauts will translocate their entire lives to the heart of the city in a 24 hours a day, non-stop, incubation event. Without a curtain in sight. Watch them eat, sleep, entertain, perform – in our very own biosphere experiment that is at the cutting edge of performance art.
>
> (Melbourne Festival Guide 1996: 38;
> cited in Rossiter and Gibson 2000: 441).

Rossiter and Gibson describe the first encounter with the Urban Design Capsule:

> When we push our way to the front of the crowd it is 'getting-ready-for-bed-time.' Some of the five bald men are in their striped pyjamas – others are still in their day suits. One is in the window/room that contains the bathroom basin, shower (with partial screen) and exercise equipment, cleaning his teeth. He turns round to the crowd, my son bears his teeth and has them scrubbed – albeit through the 'pane of separation' (Kermond 1996). The smudge of toothpaste on the inside surface remains in place all evening – a trace of the communicative act.
>
> (Rossiter and Gibson 2000: 441)

The Urban Design Capsule (UDC) experiment reveals several things about the communicative repertoire of the street. First the experiment interrupts the normal pedestrian routine on the street. People slow down or stop to stare with a sense of voyeurism and general fascination. It was estimated that over the sixteen days of the performance 200,000 people observed the spectacle with rarely fewer than 50 bystanders in front of the windows. The plate glass windows themselves put a barrier to 'normal' interaction. As Rossiter and Gibson observe, plate glass windows have been assumed to instil a 'strong sense of isolation' in the way that they divide the physical senses insulating those inside from 'the sound and touch of other human beings' outside (Sennett 1990: 109; cited in Rossiter and Gibson 2000: 443). Ironically this separation of the senses would suggest an even stronger focus on the visual, seen as the pre-eminent urban sense, especially on the street and especially as a way of limiting communication (as Simmel and Sennett observed). The experiment revealed that the window could be much more porous to acts of sociality. The situation also excludes talk. I suggest that what this does is broaden the scope of communication to include much more strongly the use of bodies. It also opens up the emotional arena. This is in part because the art-stronauts have also inverted the context in which communication is taking place. The communicative spectacle of the UDC is one of the private realm lived out in public. This seems to encourage an ethic of care and intimacy between the performers and some of the audience.

> At other times members of the crowd actively communicated with the performers. On one occasion two people shouted through the glass to Neil Thomas (who masterminded the performance), 'We've got a house-warming present for you.' They proceeded to attach a very small plant, perhaps a sweet pea, in a tiny square pot about a metre from the ground. Thomas looked truly delighted. He wrote a sign which read 'Please take care of our garden' and attached it to the inside of the window. Not long after someone else watered it. . . . The experience has been full of surprises, says Thomas. People turning up regularly with notepads to write messages on; people concerned about whether the capsulites were eating and sleeping enough; big burly guys, the type Thomas says don't usually go in for performance art, getting a charge out of the experience.
> (Schembri 1996; cited in Rossiter and Gibson 2000: 442)

There are elements of this situation that are starkly ironic. Rossiter and Gibson raise the question as to why the ethic of care extended towards the art-stronauts is not extended to the homeless who actually live their private lives in the public street and who constituted some of the audience for the UDC, strangely united in fascination with the tourists and gentrifiers of the district.

What is revealing is how the activity of the street could be transformed in this way. Rossiter and Gibson interpret this as a civility of excess 'inter-pellated as communicators across technology, physical barriers, social and cultural difference, the crowd and the UDC modelled new forms of address and care' (Rossiter and Gibson 2000: 445). For me this civility of excess is an expansion of the communicative repertoire to involve bodies and emotions. It demonstrates how the interruption of normal street etiquette, the inversion of the speech act context (private and public) and the interruption of the normal flow of communication can expand the communicative resources used in a deeper form of communicative rationality instilling an ethic of care.

PUTTING YOURSELF ON THE STREET: THE POLITICAL PROTEST OF JACKIE SMITH

On 4 April 1968, civil rights leader Martin Luther King Jr was killed on the balcony of the Lorraine Motel in Memphis, Tennessee, by a bullet fired from across the street. After a period of decline, a fund-raising effort saved the motel and converted it into a Civil Rights Museum. Across the street from the museum there is a tatty old couch on which Jacqueline Smith has conducted a continuous protest for over a decade against what one of her handmade placards calls 'The Civil Wrong Museum' (Jones 2000: 451).

As John Paul Jones III (2000) describes it, Jackie Smith's protest operates at a number of levels. She feels Dr King's philosophy of civil rights was about ongoing struggle to try to solve the root of problems. Located in a black neighbourhood with high levels of unemployment, poverty, drug abuse and crime she argues that the Lorraine Motel Museum is doing nothing to address these problems. For Jackie Smith this situation is compounded by the fact that poorer black residents are being driven out of the neighbour-hood by its gentrification, for which the Civil Rights Museum is being used as a catalyst. The Civil Rights Museum is also being used as a cultural centre where receptions and parties are held, which Jackie feels is disrespectful to the significance of the site where Martin Luther King lost his life.

Jackie Smith's street protest represents a deliberate use of presence in the street as a political act. She inverts normal communicative regimes by placing her public protest as an occupation involving her private existence (the sofa, the plant pot, the visits from neighbours and friends) in a public context. Her communication is highly situational but resonates and indeed raises an alternative voice to dominant discourses. Her protest and presence also make the point that political struggle is ongoing and should be related to con-temporary practical problems rather than the commemoration of the past in a way that might suggest that the aims of the civil rights movement had been achieved.

Jackie's street presence offers an alternative communication and reinterpretation of history. In this the street is highly symbolic; it represents for John Paul Jones

> the productive politics of this space . . . in the street that separates Jackie from the museum, that interstitial space that juxtaposes and puts into sharp relief one version of African-American history to another . . . her presence taps a surplus of meaning that exceeds the authorial inten-tionality of Lorraine's creators. It is in fact this excess that enables and makes meaningful her spatial praxis, for she demonstrates that for both

space and history, as in all things political, there was always the potential for reinterpretation, and hence always a potential oppositional moment.

(Jones 2000: 458)

Jackie Smith's bodily presence, her talk, and occupation of street space represents the communicative impact of the historical agent, the 'I' against the more institutionalized and generalized expectations of the community. Jackie represents the situational and ongoing reinterpretation of urban memory and history and, as Jones indicates, perhaps a more productive politics comes from the communicative acts of Jackie and the Civil Rights Museum, held in interpretative and situational tension on either side of the street.

What I have explored in this chapter is the way that communicative action pushes at the boundaries of ongoing interaction. Even in its most conventional manifestation as street etiquette between strangers in the city there are potential plays, unratified signals and other excesses that push for wider communication (even if they often don't succeed). Out of equilibrium communication is found in de Certeau's understanding of pedestrian rhetoric. The potential for communication on the street to act transformatively was explored by comparing Mardi Gras in New Orleans and Sydney. One festival contained the communicative potential whilst the other expanded it. Deliberate communicative disruptions can be seen in through artistic and political uses of the street (the Melbourne design capsule and the street politics of Jackie Smith) and show the way to a fuller communicative rationality. In the next chapter we consider how rationality is understood in the more intensive relationships in the neighbourhood and community.

4 In the community

INTRODUCTION

In the last chapter we saw how 'urban rationality' limited the interactions on the crowded street to the fleeting contacts of strangers. In the classic literature on urbanisation these types of distanced responses were thought to be growing within more familiar communal and neighbourly relations as well. Drawing on Tonnies' (2001) famous distinction between gemeinschaft and gesellschaft there was an extended discussion about whether the process of urbanization led to traditional communal relations (typical of village life) being replaced by ties of association. Ties of community were assumed to be rich and multi-layered, ties of association were specializsed with a more diverse range of people being known only in singular ways. There were claims for a decline of community (Wirth 1938; Stein 1960), a resurgence of community (Gans 1962; Suttles 1968) and a community of limited liability (Webber 1964; Janowitz 1967).

I begin this chapter by arguing that Tonnies' original specification of gemeinschaft and gesellschaft is as much about different forms of rationality as it is about the nature of social ties. This perspective has received much less attention in the urban literature. Tonnies makes a distinction in terms of rationality between natural will (gemeinschaft) and rational will (gesellschaft). Natural will was intuitive and organic, rational will abstract and atomistic. 'Natural will is spontaneous and unreflecting, rational will was artificial, deliberative and geared to pre-meditative "rational calculation"' (Harris 2001: xvii). I argue that this distinction between rationalities is supplanted by an understanding of 'transactional rationality' that combines intuition with abstraction, shown in the productive tension between the 'I' and the 'me' of rationality. An idea of transactional rationality I believe starts to dissolve the long-standing distinctions between the implicit and intuitive rationality of the community, and the abstract, contractual rationality of the city as a whole. The community is not defined by a coherent rationality, or moral order. There are overlapping and conflicting networks of association that exist within the community and extend beyond it. Implicit, intuitive and speculative relations can be found within community and in the city as a

whole. In all spheres there is a continuum of relations from the instrumental
and calculative, to the aesthetic and world disclosing. Crucial are the particularities of the situation in which different rationalities meet and interact. The
speculative side of rationality is suggested using Addams' (1968) reflections
on the settlement house experiment in Chicago. I then explore transactional
rationality through the spaces of performativity, discursivity and consummatory communication using Chauncey's (1994) historical study of the
emergence of gay New York.

RETHINKING GEMEINSCHAFT AND GESELLSCHAFT

We begin with Tonnies' famous distinction between gemeinschaft and gesellschaft. In the classic literature on cities this is the distinction between forms of
social interaction in different settlement types (community and association).
What I emphasize here is that fact that Tonnies was specifying different types
of rationality in the distinction between the operations of what he called
natural will and rational will. As Tonnies describes it:

> Natural or essential will is the psychological equivalent of the human
> body; it is the unifying principle of life, conceived as the pattern of
> material reality to which thinking itself belongs. . . . It involves 'thinking'
> in the sense that the organism contains certain cells in the forebrain
> which, when stimulated, cause the psychological activities that we
> interpret as thought (of which the speech faculty is undoubtedly a part).
> By contrast, rational or arbitrary or calculating will is a product of
> thought itself, and comes into being only through the agency of its
> author – the person doing the thinking – although its existence may be
> recognised and acknowledged as such by other people.
>
> (Tonnies 2001: 95–6)

The forms of will interact in many different ways but there are two primary
orientations: gemeinschaft or community and gesellschaft or association.
Gemeinschaft is the interaction of natural will and gesellschaft the association of rational will. As Harris defines it:

> In Community reason itself took the form of shared practical reason
> ('common sense' in its literal meaning), whereas in Society reason meant
> either private computation of profit and loss, or individual intellects
> grappling with 'abstract universals'. In Community, not just work but life
> itself was a 'vocation' or 'calling', whilst in Society it was like a 'business'
> organised for the attainment of some hypothetical 'happy end'.
>
> (Harris 2001: xviii)

Tonnies linked the types of will and their typical forms of interaction to
distinctive settlement forms. Gemeinschaft and natural will is of course

associated with village life and the continuity of contact and tradition that exists in such settlements. But this mutuality can also exist in the town or city-state, according to Tonnies, especially the guild cities, small self-governing urban centres. These were seen to be the highest flowering of gemeinschaft. In contrast the rational will of gesellschaft is most evident in the big city.

Drawing on the work of Dewey, Robert Park (1926) made a distinction between urban communities that possess a moral order in which the community represents collective values, and communities that result from economic competition and the sorting of occupations into neighbourhoods based on status, not solidarity. The latter owe their workings to economic rationality and often have unpredictable or unintended consequences because they are not based on a community of deliberation. The collective represent-ation of community will was for Park the moral order. For Park (1926) proper rationality is the ability to be understood in public.

The idea that rationality realizes its fullest expression in community is supported by Mead's (1934) interpretation of rationality as the response of the individual to the call of the 'generalized other'. This is the anticipation by the individual of the likely response to her/his action by the community as a whole. The ability to anticipate that response is the rational faculty. Rationality is, in this sense, the operationalization of community norms.

The interconnected and processual elements at work are reminiscent of Dewey's idea of the 'web of life', the multiple relations that organisms have with their environment (Park 1936). For human society Park makes a distinction between the biotic and the cultural 'the web of communication which man has spread over the earth is something different from the "web of life" which binds living creatures all over the world in a vital nexus' (Park 1936: 12). And elsewhere: 'There is a symbiotic society based on competition and a cultural society based on communication and consensus' (Park 1936: 13). The biotic realm is governed by competition or the struggle for life, the cultural by self-restraint and consensus. The whole system is characterised by competitive cooperation.

There are intimations here of a division in thinking that has structured thinking about cities and the public realm. Habermas (1984, 1987) makes a distinction between instrumental and communicative rationality. The first is the rationality of capitalist economics, which structures the system. Com-municative rationality is, on the other hand, that of the lifeworld, of every-day speech that is not seeking success by competition but cooperation, and at its root cooperation as the communication of meaning.

The assumptions of the instrumental view of rationality in the reproduc-tion of community can be seen in the work of the early Chicago School. The idea of community from the Chicago School allied with a heavily socialized interpretation in the work of the symbolic interactionists which meant that the pragmatic inheritance in urban studies was separated from some of its more naturalistic roots. Interaction was concerned mostly with spoken inter-action between people. Language was the epitome of a symbol-producing system. Symbolic interaction was privileged over non-symbolic interaction because it was over linguistic symbols that a distinctive human capacity to respond to the 'generalized other' was instilled. That is, the individual is able to anticipate and take in the response to a symbol as if the whole community were responding. Symbols are signs of common responses. The symbolic process is reflexive – in using the symbol a person stimulates himself as he stimulates others in a given situation. As Kang (1976) argues, here the individual takes himself to be an object (or experiences himself as an object) in the way the (specific) others respond to him as an object (or experience him as an object). In this sense the individual takes the role of the others. It is also regulative:

> if in using a symbol the individual stimulates the others and himself as he would stimulate any individual (in the same way) in the situation. Here the individual takes himself to be an object in the way any other individual would respond to him as an object.
>
> (Kang 1976: 130)

This is the role of the generalized other. This symbolic process is like '(1) looking at oneself in a mirror made of others and (2) improvising one's acts in the world (or theatre) of this mirror in search of repeatable working roles' (Kang 1976: 130–1). If participants in this form of interaction have to take themselves as objects in order to communicate, that has the effect of instrumentalizsing these relations within community norms.

Although the work of the early Chicago School has been critiqued for its implicit white, middle-class patrician viewpoint, there is another way that the investigation of 'deviant' communities can be read. Beginning with Jackson and Smith's (1984) more sympathetic treatment of this work we can view it as an early encounter with difference (see Pratt 1998). In fact the 'mosaic of little worlds' that Park (1969) and others observed in Chicago was a testament to the way that a range of social conventions can develop in an urban context. Individuals take on the generalized other of community norms (in this case in the anticipation of the likely response to an action) and in so doing reproduce that norm. The logic of community reproduction can indeed overwhelm other sources of difference within the community.

Solidarity with community can be seen in this way as a product of power. As Hoch (1996: 36) puts it: 'Communities are shaped not solely by mutual interdependence among members but also by the imposition of power relations that undermine or sustain this solidarity.'

In many cases community norms are reproduced by its members even if those members' interests are not served by them. Bourdieu (1984) pointed to one mechanism for this – the tacit reproduction of dispositions on the body in social space. But these burdens can weigh on the mind directly. Understandings of cognitive rationality from game theory suggest that conventions may be desirable or undesirable to people within the community as well as without, but that they get reproduced simply because of the weight of anticipated expectation that accompanies routine actions and communications within the community. Given what the individual expects others to expect, subsequent actions will be severely circumscribed. These are critical elements in the emergence and reproduction of norms (see especially Ullman-Margalit 1977). Of course community expectations are what enable anyone to act in the first place. Dewey (1922) recognized this is his characterization of habit as an active force, an exercise of will, rather than some form of passive momentum.

To what extent then is rationality the keeper of community boundaries in the circulation of power? To what extent is the city destined to continue as an archipelago of little worlds, each defined by the ongoing rationality, which homogenizes community members and limits their possibilities to those of the community? We have seen how these barriers can be maintained by body movement and in encounters in the street and also how these distinctions can be broken down by unusual events and by an underlying ritual of respect in interaction. Surely though, the hold of rational expectations is all the greater in community, implicated as it is with community norms and accumulated expectations in the history of the community.

Yet the city is not just a settled archipelago of separate islands of distinct communities. There are instances where norms are being challenged or changed or are becoming incoherent. More and more there are examples of city residents with multiple loyalties to many different communities. There is also a growing appreciation that identities are often not defined by singular associations, or histories, but are increasingly hybrid.

COMMUNITY AND HYBRIDITY

People can occupy multiple communities or parts of communities. They live in networks of association, some overlapping, some separate. They might be subject to competing or contrasting expectations in the different social or spatial realms. Actors can play different roles in the different networks. Goffman (1974) captures some of this in his 'frame analysis', by which the framing of the situation gives the cue for interpretation. Membership of a range of different networks might be assumed to allow a greater degree of autonomy, a freedom of movement resulting from the fact that no one network of social realm has exclusive influence over the actor. Indeed ideas of cosmopolitanism have traditionally built on this assumption (discussed in Chapter 8). The degree of autonomy in multiple networks is on effect of power however. As I have argued earlier (Bridge 1997b) if the actor is simply entrained in a number of different networks then the sense of self might be torn apart – they might literally experience the sense of time-space compression (Harvey 1989). Equally, confinement within a single network can lead to an experience of enclosure within an ever-present space–time.

The fact that the multiple spaces and associations with which the subject must identify leads to a more complex hybrid identity (Tajbakhsh 2001). Where networks provide opportunities rather than enclosures then the actor may experience an expanded reach of space-time and an enhanced sense of self. A sense of self is interpreted by White (1992) in network terms as 'personhood.' Personhood is a network effect that results not from stability and centrality of experience but movement around a range of overlapping networks. As White expresses it: 'Persons come to be generated only out of large-scale frictions amongst distinct networks populations' (White 1992: 197). Thus personhood reduces the ambiguity of social life to sustain orderly interpretative frames at a cost of increased ambage (circuitousness of networks). White goes on to say that 'stable identities as persons are difficult to build; they are achieved only in some social contexts, they are not pre-given analytic foci'.

There is a range of transactions in an array of networks. 'The imaginaries most often employed in speaking of transactions are accordingly those of complex joint activity, in which it makes no sense to envision constituent elements apart from the flows within which they are involved (and vice versa)' (Emirbayer 1997: 289). As Dewey argues 'the import of . . . essences is the consequence of social interactions, of companionship, mutual assistance, direction and concerted action in fighting, festivity, and work' (Dewey 1929 142, cited in Emirbayer 289). 'No one would be able successfully to speak of the hunt*er* and the hunt*ed* as isolated with respect to hunt*ing*. Yet it is just as absurd to set up hunting as an event in isolation from the spatio- temporal connection of all the components.' (Dewey and Bentley 1991, cited in Emirbayer 1997: 289).

In their contemporary interpretation of transaction Emirbayer and Mische (1998) use these insights to reconstruct the idea of agency as

> temporally constructed engagements by actors of different structural environments – the temporal relational contexts of action – which, through the interplay of habit, imagination, and judgement, both repro- duces and transforms those structures in interactive response to the problems posed by changing historical situations.
>
> (Emirbayer and Mische 1998: 970)

This involves three constitutive elements of agency: iteration, projectivity and practical evaluation. This is a historically sensitive view of agency with variable reflexivity and it is a view of agency that is inherently social and relational. In fact they categorize the whole approach as one of 'relational pragmatics'.

The focus on transaction in relational pragmatics allies strongly with research methods in which interaction and relation is central, most pro- minently social network analysis. Indeed network analysis, from these traditional sociological investigations to actor network theory, has become the privileged metaphor and tool for understanding social and economic change in late modernity (such as in Castells' trilogy on *The Information Age* 1996, 1997: 98). Rather than relying on traditional assumptions about solidarities of community or society, network structures and their relations might help us capture more adequately the realities of everyday life.

I have argued elsewhere that networks are invaluable in understanding the linkage between power and the experience of space–time (Bridge 1997b). 'Actors who are positioned at the intersection of multiple temporal-relational context can develop greater capacities for creative and critical intervention' (Emirbayer and Mische 1998: 1007).

In the contemporary city these transactional environments are becoming ever more complex: the range of media through which communication can be made and the diversity of groups and situations to which they link. 'Social networks cut across discrete communities and other entities and are interstitial, even though in certain cases they may also congeal into bounded groups and clusters' (Emirbayer 1997: 299). These overlapping power networks have normative effects. 'For pragmatists normative implications flow naturally out of the central concept of transaction itself' (Emirbayer 1997: 309):

> Thus values are by-products of actors' engagements with one another in ambiguous and challenging circumstances, which emerge when individuals experience a discordance between the claims of multiple normative commitments. Problematic situations of this sort become resolved only when actors reconstruct the relational contexts within which they are embedded, and in the process, transform their own values and themselves.
>
> (Emirbayer 1997: 310)

'The appearance of . . . different interests in the forum of reflection [leads to] the reconstruction of the social world, and the consequent appearance of the new self that answers to the new object' (Mead 1964: 149; cited in Emirbayer 1997: 310). Evaluation is a result of transaction (Joas 1985). Rationality is deeply implicated in these hybrid situations.

As the book has suggested so far, intelligence is not just confined to the calculation of instrumental interests. Non-discursive and implicit communications are endemic to communal being. This form of community problem solving provides us with the possibility of a leading edge of rationality that is tied as much to the speculative activity that can enhance community. I think there are also suggestions of the breakdown of oppositions between community 'me' and individual 'I' or between an intuitive and spontaneous rationality on the one hand and an abstract and atomized rationality on the other. It is a rationality that is neither the intuition of gemeinschaft nor the calculation of gesellschaft, not a rationality within community or outside communal interests but a blurring of these boundaries in the idea of transactional rationality.

TRANSACTIONAL RATIONALITY AND COMMUNITY

Communicative rationality for Habermas is conformity to Mead's 'me', meeting the expectations of the generalized other through expressions of objective truth, normative rightness and subjective sincerity. The good reasons or grounds for these statements examined in debate are what constitute a substantive notion of rationality. Dewey however had an idea of rationality that was the mediation of the 'I' and the 'me'. Dewey called the 'I' 'impulsion'. It is speculative, unpredictable, often at large in displacements, or miscues, or at the edges of communication. It embraces an expanded idea of communication involving discursive and non-discursive, performative as well as deliberative elements, and aesthetic communication. As I discussed in Chapter 2, aesthetic experience can be seen as part of everyday life, an ongoing experience if we take Dewey's (1987) emphasis on art*work*, on the activity of producing art rather than the art object itself, of which on the latter most aesthetic theory concentrates.

The idea of communication as constitutive rather than representational aligns with an understanding of the world disclosing aspects of aesthetic experience. Whereas Habermas believes aesthetic world disclosure cannot be encompassed within the realm of rational argumentation, Wellmer (1991) and others, drawing in part upon Dewey, as well as Derrida and Heidegger, suggest how world disclosing language is vital for intellectual and moral progress. As Duvenage (2003: 132) argues 'any theory of rationality that fails to incorporate a positive account of novel experience and creative meaning change is inadequate'. Habermas considers only the rationality of validity-oriented speech and action (the rationality of reason giving), not the rationality associated with disclosing different horizons of meaning. For Dewey the activity of disclosure is one type of action on a cognitive continuum. Dewey does not draw 'a too strong cognitive boundary between the capacity for world disclosing and the capacity for giving reasons' (Kompridis, cited in Duvenage 2003: 134).

'Social change is predicated here not just on the linguistically mediated "me" as it makes an appearance in discursive communication but on the instinctive, aesthetic, unpremeditated "I" that bursts forth on the social scene and makes individual experience valuable to the community as a whole' (Shalin 1992: 271).

> This universe is composed of many verses and is shot through with competing perspectives. It allows reason to be scattered across disparate social niches: it makes it appear under jarring sexual, racial, ethnic, religious, cultural and social guises; it does not demand that various life-forms be brought to a common denominator other than their proponents' commitment to coexist peacefully, respect each other's uniqueness, and, where possible, draw on experience accumulated by others. As such,

the pluralistic universe serves as the epitome of modernity pragmatically understood . . . reason uncaged is reason enlightened by sentiment, sensitised to uncertainty, steeped in ambivalence, humbled by the consciousness of the limits that nature sets on its ambitions.

(Shalin 1992: 272, 274)

Reason is also ever-expansive beyond these boundaries in a number of ways. Rorty's (2000) limiting of rationality to community is in part explained by the assumption that discourse encompasses experience and reason. Yet for Dewey and James aesthetic and religious experience was consummatory and explosive and not explicable in language:

a universe of experience is a precondition of a universe of discourse. Without its controlling presence, there is no way to determine the relevance, weight, or coherence, of any designated distinction or relation. The universe of experience surrounds and regulates the universe of discourse but never appears as such within the latter. Although lived experience may exceed the boundaries of discourse, our expression of it usually, and our discussion of it always, cannot.

(Dewey 1938: 74, 100 [. . .])

Such world disclosure has decentring and deconstructive dimensions. Dewey argues that these forces are linked to a reconstructive moment:

disclosure . . . of possibilities that contrast with actual conditions . . . are . . . the most penetrating 'criticism' of the latter that can be made. It is by a sense of possibilities opening up before us that we become aware of the constrictions that hem us and of burdens that oppress.

(Dewey 1987, cited in Duvenage 2003: 135)

As Duvenage argues, 'the decentring effects of disclosure can be handled properly only through the constant activity of reconstructing shattered interpretations of the world in the light of new ones' (Duvenage 2003: 135). This leads to an idea of plural rationalities and a fallibilistic notion of truth. The plurality of rationalities, for Wellmer and Dewey, seeks to mediate between world disclosure and discursive reason (Duvenage 20003: 137). Aesthetic validity claims draw on local context but can be deployed in discursive reason.

TRANSACTIONAL RATIONALITY AND THE CITY

Broadening the idea of communication and dissolving the distinction between instrumental and communicative rationality changes our view of the community and the city in very striking ways. First we have a view of language as constitutive, rather than representative. That has certainly been part of the Chicago School and ethnomethodological legacy. It can be broadened to take on non-symbolic, non-discursive communication. It also allows a fuller engagement with ideas of power and the more macro, structural issues that critics have argued was missing from symbolic interactionism. The performativity of everyday life can be considered not as the interaction of pre-constituted atomistic agents, but rather transaction and communication. This suggests how meanings can be forged in the interstices of power but also how communicative competences reproduce power. Actual communicative activity has both instrumental action-oriented and meaning-oriented goals – its goals are both constitutive and representational. What this means is that power and the potential resistance to power are ingrained in the communicative repertoire rather than the separation of a lifeworld full of meaning and a system full of power. As Langsdorf puts it:

> In practice, this means that participants are at least as much engaged in constituting what Foucault (1972, 37) calls 'discursive formations' or 'systems of dispersion' as in reaching understanding. What is 'dispersed' as aspects of a cultural scene are differently interpreted by its inhabitants as meaning. This dispersal has perlocutionary effect – including, the ongoing constitution of participants who are 'actors' seeking 'success'. Particularly in oppressive socio-political contexts, when a non-engaged observer (analyst or critic) may see no means for altering the cultural scene, the 'looking differently' that is phenomenology can discern participants engaged in strategic communicative action that realigns their identifications with aspects of their cultural scene. In doing that, they actualise their identities differently – and so, succeed in constituting themselves differently. Often, given the recalcitrant power of oppressive cultural forms, they are careful to conceal that success.
>
> (Langsdorf 2000: 42)

Communicative tensions were also at the heart of Walter Benjamin's (1999) arguments about how aesthetics had emancipatory potential but were increasingly commodified in the modern metropolis. Benjamin suggests the world-disclosing potential of aesthetic experiences of the city. As Hannah Arendt observed, he is able to reveal the strange correlation 'between a street scene, a speculation on the stock exchange, a poem, a thought, with the hidden line which holds them together and enables the historian or philologist to recognise that they must all be placed in the same period' (Arendt 1968, cited in Duvenage 2003: 47). Benjamin evokes the dreamlike quality of the city, which the inhabitants can move through as though sleepwalking. This dreamlike existence is enhanced by the commercial elements of the modern city such as the famous arcades of Paris – enclosed passages for shopping which created a synaesthesia of sights and sound that seduced the consumer. In Buck-Morss's evocative analysis:

constructed like a church in the shape of a cross . . . these privately owned, publicly traversed passages displayed commodities in the window showcases like icons in niches. The very profane pleasure houses found there tempted passersby with gastronomical perfections, intoxicating drinks, wealth without labour at the roulette wheel, gaiety in the vaudeville theatres, and, in the first-floor galleries, transports of sexual pleasure sold by a heavenly host of fashionably dressed ladies of the night.

(Buck-Morss 1989: 83)

This was the urban phantasmagoria that extended across the city as a whole and was found in the Great Exhibitions and World Fairs. It was the emblem of modern urbanism. This combined the aesthetic power and labyrinthine nature of the modern city in a way that Lash (1999) argues resonates with another perspective on modernity and 'a different rationality'.

As Pile (2000) explains, Benjamin believed that dreams were most vivid at the point of waking and this means that Benjamin was most interested in those parts of the city that were being torn down or radically altered. These were the places that would shake the inhabitants out of their consumerist slumber and see the city anew. It was a form of world disclosure. 'If he could bring the pieces into tension, through "dialectical imaging" by putting the pieces side by side, Benjamin thought it would be possible to induce a shock that would wake up the moderns' (Pile 2000: 79).

Rather than setting aesthetic experience aside from rational deliberation Dewey's approach, and one greatly developed by contemporary movements in the philosophy of communication, sees rationality as a form of social intelligence that seeks to mediate between aesthetic world disclosure and argumentation. Indeed world-disclosing events and effects can be seen as the leading edge of rationality, its speculative, experimental side (see Rosenthal 1990, 2002). That is why aesthetic revelation has often been associated with subjective, individual experience as the historical 'I' makes its effects on the 'generalized other' of the community. However Dewey's work, as Wellmer and Duvenage reveal, does not necessitate this connection. Indeed aesthetics here is seen as another form of communication, usually involving contexts and contents that are impossible to represent in language.

The different strands of communication, from the instrumental to the aesthetic, and the blurring of the boundaries of reason in the idea of transactional rationality can be seen in various studies of the city. These trends can often be mostly clearly seen in an emerging community that is struggling for identity. They were shown in the experience of Jane Addams and the other residents of Hull House in Chicago in the late nineteenth and early twentieth centuries – the classic example of social interventions to try to assist newly arrived immigrant communities in the city. Hull House was a settlement house in a poor neighbourhood in Chicago where a number of philanthropic middle class women lived in an attempt to help the people in the neighbourhood deal with the social problems of the neighbourhood. It was one of the early examples of a social movement to change the conditions of the urban poor that spread across US cities. The middle class women of Hull House showed an ability to learn from the people they were trying to assist, to take a practical, learning-by-doing approach. Much of what they did was experimental and speculative. Hull House was a practical attempt to help build community in conditions of ethnic and social diversity and deprivation. It is, I suggest, indicative of a wider transactional rationality.

HULL HOUSE: THE EARLY EXAMPLE OF AN EMERGING TRANSACTIONAL RATIONALITY

Hull House was an early example of a social settlement in which middle class philanthropists and intellectuals lived in a deprived neighbourhood and sought to help through education programmes and nascent forms of social work. There were strong links between John Dewey in his time at the University of Chicago (1894–1904) and the Hull House settlement. He was a friend of Jane Addams (who founded Hull House) and a member of the board of trustees. The practical activities of Hull House residents in trying to facilitate change in the neighbourhood, as well as the intellectual reflections of Jane Addams and others, deeply influenced Dewey's views on education and his philosophy more widely (Siegfried 1996: 73–9). What were initially vague aspirations for curing the ills of their society took on the definite structure of experimental projects at Hull House. They did not have to appeal to tradition, which was particularly useful given the constraints on the Hull House women in a patriarchal society. Hull House was a community in itself but also a laboratory for more extensive community building. Describing Jane Addams' efforts Siegfried suggests that 'she sought to build on the fact that real cooperation was possible among the diverse immigrant populations in the Hull House neighbourhood and looked for ways to extend the mechanisms of cooperation back to their mutually hostile countries of origin' (Siegfried 1996: 75)

Siegfried's depiction of Addams' conduct at Hull House I argue epitomizes an emergent transactional rationality, beginning with the speculative and concrete encounter with the other leading to more definite understandings that promoted cooperative activity, criticism of inherited structures and the transformation of possibilities to help create a better future.

> Jane Addams' insistence that reciprocity ought to characterise the relationship between social worker and client, teacher and student, and her testimony that she daily learned as much from the poor among whom she chose to live as she taught them, were neither a priori deductions from moral principles nor idle platitudes, but conclusions reached from reflections on her experience. She consciously tested her own beliefs in her interactions with others and discarded or revised them as needed. She brought with her to Hull House a profound respect for the

Other in her or his uniqueness, for example, but throughout *Twenty Years at Hull-House* she reveals how the concrete specificity of her interactions helped her recognise the class and ethnic prejudices informing her good intentions while at the same time providing the means for developing an authentic appreciation of and a more knowledgeable response to a lived diversity that could not have been predicted beforehand.

(Siegfried 1996: 78)

In a paper written before settling at Hull House Addams expressed her political perspective:

> . . . in a democratic country nothing can be permanently achieved save through the masses of the people, it will be impossible to establish a higher political life than the people themselves crave; that it is difficult to see how the notion of a higher civic life can be fostered save through common intercourse; that the blessings we associate with a life of refinement and cultivation can be made universal and must be made universal if they are to be made permanent; that the good we secure for ourselves is precarious and uncertain, is floating in mid-air, until it is secured for all of us and incorporated into our common life.
>
> (Addams 1968: 116)

Addams describes the activities at Hull House, which consisted of a series of experiments in social cooperation: small-scale, pragmatic and problem oriented. For instance the Cooperative Coal Association flourished for three years in helping to provide people with sufficient fuel but failed on the overly philanthropic disbursement of coal. The speculative, spontaneous nature of the initiatives often explained their success (as well as failure). After the coal coop Addams goes on to describe another initiative.

> Our next cooperative experiment was much more successful, perhaps because it was much more spontaneous. At a meeting of working girls held at Hull-House during a strike in a large shoe factory, the discussions make it clear that the strikers who had been most easily frightened, and therefore first to capitulate, were naturally those girls who were paying board and were afraid of being put out if they fell too far behind. After a recital of a case of particular hardship one of them exclaimed: 'Wouldn't it be fine if we had a boarding club of our own, and then we could stand by each other in a time like this?' After that events moved quickly. We read aloud together Beatrice Potter's little book on 'Cooperation,' and discussed all the details and fascinations of such an undertaking, and on the first of May, 1891, two comfortable apartments near Hull-House were rented and furnished. The settlement was responsible for the furniture and paid the first month's rent, but beyond that the members managed the club themselves. The undertaking

'marched' as the French say, from the very first, and always on its own feet. Although there were difficulties, none of them proved insurmountable, which was a matter for great satisfaction in the face of a statement made by the head of the United States Department of Labor, who, on a visit to the club when it was but two years old, said that his department had investigated may cooperative undertakings, and that none founded and managed by women had ever succeeded. At the end of the third year the club occupied all of the six apartments which the original building contained, and numbered fifty members.

(Addams 1968: 135–7)

Addams' account of life at Hull House is full of these examples of speculative, transactional rationality.

The speculative side of rationality is also made plain when the norms of that community contrast strongly with the norms of the dominant community and when the emerging community is resisting oppression. Here I use the example provided by Manual Castells (1983) on San Francisco in the 1970s and in particular by George Chauncey (1994) of the emergence of the gay community in Greenwich Village in Manhattan in the late nineteenth and early twentieth centuries. Chauncey's magisterial study I think is strongly suggestive of an emergent transactional rationality in the city. The study reveals the porosity of community and the situational significance of the proximity to different community norms, which mutually condition the types of gay identity that emerge. Chauncey shows how identities are performative and transactional. His study is a pre-eminent example of the situational character of emerging identities, forms of communication and community norms and as such I consider this work in some depth.

TRANSACTIONAL SEXUALITIES IN SAN FRANCISCO AND NEW YORK

If we look at the literature on queer space we see many of the elements of an emerging transactional rationality. The significance of 'the situation' and how it can exceed its discursive confines is evident in some of the happenings that help establish a space for gayness. In his discussion of the gay community in San Francisco in the 1970s, Castells (1983) points to the significance of a waiter/drag queen Jose Sarria at the Black Cat bar. One night Sarria gave a rendition that portrayed Carmen as a drag queen hiding from the police in the bushes of Union Square (a fashionable public square in the centre of downtown San Francisco). The performance was so successful that Sarria was hired by the straight owner of the Black Cat and performed to packed audiences on Sunday afternoons. As Castells notes, the bars and the drag queens were fundamental to the creation of networks and making gay people visible. The key movement was from the bars to the street. What Castells points to is the fact that gays were able to negotiate their emergence from a covert existence into an open alternative lifestyle using Beatnik culture and its experimental elements. The alternative urban space of the North Beach area provided the more fluid environment in which the speculative communicative acts of a group trying to establish itself as a community could be transacted. At this point the emphasis was more on the experimental 'I' of communication rather than a rigid adherence to the community as a 'me' that must accompany all my speech acts.

The emergence of gay identity and recognizable space is also evident in George Chauncey's fascinating account of gay New York, 1890–1930. Rather than being structured according to the divisions between homosexuality and heterosexuality, as they were after the Second World War, Chauncey's detailed account shows the porous nature of sexual identities in the modern city.

Communicative spaces: performativity

Chauncey's study suggests that sexual identities were situational and trans-actional, rather than essentialized, at this time. As Chauncey puts it:

> The most striking difference between the dominant sexual culture in the early twentieth century and that of our own era is the degree to which the earlier culture permitted men to engage in sexual relations with other men, often on a regular basis, without requiring them to regard them-selves – or to be regarded by others – as gay ... the abnormality (or 'queerness') of the 'fairy' ... was defined as much by his 'woman like' character of 'effeminacy' as his solicitation of male sexual partners; the 'man' who responded to his solicitations – no matter how often – was not considered abnormal, a homosexual, so long as he abided by masculine gender conventions. Indeed, the centrality of effeminacy to their representation of the 'fairy' allowed many conventionally masculine men, especially unmarried men living in sex-segregated immigrant communities, to engage in extensive sexual activity with other men without risking stigmatisation and the loss of their status as 'normal men'.
>
> (Chauncey 1994: 13)

In the working class 'Bachelor sub-culture' which included young unmarried seamen and transient labourers, the status of manliness was something that was achieved, and could be lost. It was 'a kind of ongoing performance' (Chauncey 1994: 80).

The fluidity of sexual identity was also a result of a spectrum of ethnic and regional attitudes and practices in the immigrant communities that bounded each other. A study in 1921 found that in the Italian neighbour-hood of the Lower East Side there were numerous fairy saloons interspersed with saloons where the female prostitutes worked (Chauncey 1994: 72). In the Jewish neighbourhood of the Lower East Side there were no open 'fairy resorts', although there were numerous tenements and street corners where female prostitutes worked. In other locations there was less distinction and greater fluidity to sexual practices between ethnic groups, especially in the male immigrant 'bachelor sub-culture' that involved Italian, Black and Anglo-American men housed in the tenements of the Bowery and elsewhere.

This socially constitutive and ongoing nature of sexual identity was investigated by labelling and role theory in the symbolic interactionist tradition (see especially McIntosh 1968; Gagnon and Simon 1974; Plummer 1975). As Plummer argues (1975: 30) the fundamental axiom of the inter-actionist approach to sexual identity is that 'nothing is sexual but naming makes it so. Sexuality is a social construction learnt in interaction with others'. At the time Plummer rendered his interactionist account of sexual stigma, homosexuality was seen in terms of the sociological category of deviancy (that is non-conventional behaviour) and it registered the inter-actional effects of labelling by wider society and how that was internalized through self-reflexive control that often demanded that the gay person 'passed' as straight.

Communicative spaces: discursivity

Frank Mort (2000) argues that the representational qualities of sexualities were anticipated in interactionist investigations. As Mort (2000: 309) argues, the interactionist focus on roles and labelling and the socially constructed nature of sexual acts and values anticipated Foucault's (1978) insights into the representational quality of sexuality. Foucault's arguments concerning the production of sexuality via discourse in a range of sites became highly influential. The power discourse of sexuality was dispersed and decentred, in a range of settings, from sanitary schemes for urban improvement, house-hold manuals, statistical tables, medical dossiers and census returns. This discourse operated through and beyond the sorts of interactive settings that Plummer had identified. The discourse of sexuality certainly helped define the situation in which partners in interaction found themselves.

Mort (2000: 307) gives an example of a discursive complex that occurred in London in 1953, the year of Queen Elizabeth II's coronation. London was the focus of national and Commonwealth attention and some of its more transgressive activities came under scrutiny. The tabloid press ran sensation-alist headlines on the twin vices of male homosexuality and female prostitu-tion in districts adjacent to the coronation route, especially Soho. In the following year the Conservative Government commissioned the Home Office Committee on Homosexual Offences and Prostitution (The Wolfenden Committee). This committee recommended the partial decriminalizsation of male homosexuality but also sought to regulate homosexuality and prostitu-tion by making them visible and engaging on extensive debate on the problems (in a way reminiscent of Foucault's analysis; Mort 2000). Central London was mapped in term of irregular sexualities.

The Metropolitan police commissioner submitted a document which was, in effect a homosexual map of London. Spanning the whole of the greater West End, it ranged from Kensington Gardens, Knightsbridge and Hyde Park, through Victoria and across Bloomsbury and the Strand. Circled in red were the places where police arrests for importuning and gross indecency were most frequently made. The familiar landmarks of London's central districts had been redrawn to reveal the spaces of homosexual desire.

(Mort 2000: 308)

Similar examples of the official discourse of sexuality are revealed in the medical reports on homosexuality that appeared from the late nineteenth century onwards. However, Chauncey (1994) asserts in the case of New York that the significance of such reports had been over-estimated. They are only one in a range of writings and analyses. The influence of medical reports became significant only from the mid-twentieth century, according to Chauncey. Before that:

while a few boys were diagnosed as homosexuals by doctors, many more were denounced as queers by other boys on their street. Most men who escaped such denunciations did not begin to think they were fairies because they read about them in articles published in obscure medical journals, but because they met fairies in the streets and were confronted every day by the inconsistency between their desires and those pro-claimed by the men and women around them. The fairy's positions in the sex-gender system made sense to them not because it had been constructed (or explained) so carefully by elite writers, but because it seemed reasonable in terms of the social practices that constituted and reconstituted gender on an everyday basis.

(Chauncey 1994: 125)

Chauncey emphasizes the importance of ongoing, everyday, situated, com-municative practices in the forging of homosexual identity. Gay subjectivity was not produced in professional discourses but via situational encounters and influences. There were a range of sites, performative and discursive, in which various communications and miscommunications over elements of gay-ness were transacted. Micro-geographic and performative variations continued to constitute gayness in Greenwich Village and Harlem. The community was more fragmented than some of its discursive manifestations suggested. There was a spectrum of identities that made up gay New York. That variation, or difference, was embedded in the emerging discursive regimes.

The forging of a gay identity did not just rely on its interactive situation but started to resonate in terms of dispersed discourse, an emerging articul-ation and argumentation over gayness. This happened in a diversity of urban spaces, both performative and discursive. Fairies made their point through visible effeminacy and the subversion of straight gendered street interaction by attempting to limit straight men's ridicule of them. Chauncey describes how a number of lesbians and gay men 'considered it strategically useful' to state their case to the famous sex reformer Dr William J. Robinson. Denying that their condition had anything to do with degeneracy or anxiety these patients claimed that 'they stand on a higher level than those normally sexed, that they are the specially favoured of the muses of poetry and the arts'. Although he ridiculed their claims Robinson did admit that these conversations had led him to take a 'broader, more tolerant, perhaps even more sympathetic [attitude]' (Robinson, cited in Chauncey 1994: 282). Letters to the editors of unsympathetic local newspapers also helped disperse the argument.

Chauncey argues that the creation of a gay folklore also dispersed the discourse and helped build up a collective identity. Through interviews with medics, conversations and in novels and other writings there was a move to historical reclamation by claiming respectable historical figures such as Shakespeare, Michelangelo and Walt Whitman as gay but also through reference to gay identity in other societies, either in history (ancient Greece) or elsewhere (via anthropological texts). These interventions stressed the nobility of male affection and love and of course they were highly contested by wider society.

Another key component in the dispersal of discourse in a hostile environment was the creation of a gay language, one that played on conventional meanings in a form of double entendre. Terms were adapted from the slang of female prostitutes.

> Gay itself referred to female prostitutes before it referred to gay men; trade and trick referred to prostitute's customers before they referred to gay men's partners; and cruising referred to a street walker's search for partner before it referred to a gay man's. Other terms, such as coming out, burlesqued the rituals of society women.
>
> (Chauncey 1994: 287)

The use of gay argot helped create a collective identity and a sense of being an insider rather than the more usual experience of being an outsider. This cultural strategy, like others, 'offered practical assistance in dealing with a hostile world', and 'such strategies also affirmed . . . cultural distinctiveness and solidarity' (Chauncey 1994: 287). The argot allowed gay men to communicate within the straight world. In many senses gay men were pre-eminent users of codes and symbols.

> The homosexual . . . is a prodigious consumer of signs – of hidden meanings, hidden systems, hidden potentiality. Exclusion from the common code impels the frenzied quest: the momentary glimpse, the scrambled figure, the chance encounter, the reverse image, the sudden slippage, the lowered guard.
>
> (Beaver 1981, cited in Chauncey 1994: 287)

Gay men appropriated and inverted words to do with 'feminine' behaviour and family mores.

Gay men used a range of tactics that played on the conventions of street interaction in order to communicate and make contact with each other, such as returning a look between men that would be cut short between straight men, or asking for the time or for a match. The emergence of elaborate window displays in the burgeoning department stores of New York at this time (late nineteenth early twentieth centuries) enabled gay men to mimic female street sociability by loitering to look at the displays and to exchange comments as a way of assessing whether they wanted to continue the contact. The anonymity of busy 'entertainment' streets in theatreland was a resource, as were certain bars, speakeasies and restaurants. And of course public parks where gay men could cruise without appearing too conspicuous were important meeting places. What Chauncey conveys is just how co-extensive gay New York was, co-inhabiting (with varying degrees of invisibility), the spaces and neighbourhoods of the dominant and often hostile social conventions. They were elaborate and extensive networks of contact, rather than the distinct mosaics of little worlds of distinctive community that Park spoke of. As well as this co-extensive environment, with various degrees of secrecy, there were extraordinary spaces that allowed communication as consummation of gay lives: the drag balls.

Communicative spaces: consummatory events

As Chauncey describes, the pre-eminent spaces of collective gay identity in New York (and Chicago and New Orleans) were the drag balls. Like many of the gay sub-cultural codes, drag balls drew on and inverted the norms of dominant society. Masquerade balls, where normal roles were inverted, and different classes mixed and danced with each other behind masks, had centuries-long tradition. At drag (or transvestite) balls men dressed as women, danced with each other and participated in fashion competitions. Subject of course to police raids and various forms of repression, the diverse range of balls at different venues in New York represented a key gay space, with high visibility and a celebratory atmosphere that ball goers sometimes took out into the streets and restaurants where they were again rebuffed by the prevailing norms of the dominant society. But within the space of the ballroom there was created a consummatory gay space that existed over and above the separate social networks of gay life.

> Part of the attraction of an amusement district such as Greenwich Village, like that of Harlem, was that it constituted a liminal space where visitors were encouraged to disregard some of the social injunctions that normally constrained their behaviour, where they could observe and vicariously experience behaviour that in other settings – particularly their own neighbourhoods – they might consider objectionable enough

to suppress. The organisers of the balls were well aware of this phenomenon and welcomed the presence of flamboyant gay men – sometimes
making them a part of the pageants they staged – precisely because they
know they enhanced the reputation and appeal of such events. As the
Liberal Club's Golden Ball of Isis, attended by two thousand people in
February 1917, Horace Mann, (well known to the audience, apparently,
as a 'noted homosexualist') took the major role of the slave in love with
the Egyptian goddess Isis in the 1 a.m. pageant.

(Chauncey 1994: 236–7)

As well as variations of sexual identity (including fairies, queers, gays, wolves
and punks) and the inflections by class and ethnicity, there were diversities
that arose from particular clubs and streets as well as between whole districts. In the latter regard Harlem had an even livelier gay life than Greenwich
Village, although this identity was subsumed under its pre-eminent reputation as the black metropolis.

Chauncey's work on the emergence of the gay community in New York
demonstrates that different spaces of communication – performative, discursive and consummatory – can all play a role in the forging of identity and
recognition. These spaces are not confined to lifeworlds or safe communities
where they can find respite from power, but rather are co-extensive with
power, conflict and social juxtaposition. They are co-extensive with the city
as a whole. We now take these insights to explore how far these alternative
communicative spaces of rationality can offer a renewed public realm.

5 In the public realm

Can the public spaces of the city nurture any kind of wider public realm, beyond individual self-interest or community norms? And if so, can we have a public realm that accommodates difference? The character of public space in cities and the possibilities of a self-constituting public have long fascinated urban scholars and philosophers. These are ever more pressing questions given the segregated and competitive nature of many of our cities. Recent philosophical understandings of the public and the city have offered a number of possibilities. There is a form of cognitive communicative rationality that transcends difference (Habermas), an emotional urban performativity that dramatizes difference (Sennett) and a kind of urban encounter between strangers, the particularity of which makes difference undeniable (Young).

Although communicative, performative and undeniable encounters are part of everyday life, in these conceptions of the public they all rely strongly on set-piece interactions that are focused and absorbing. They are deliberately meaningful in terms of communication. They also depend on a relatively static, or at least 'stilled' idea of urban space. What I explore in this chapter is an understanding of everyday life and the public that comes from the edges of communication (rather than the centre of communication) and involves all the improvisations and non-discursive communications that go on through bodies, in tacit routines in the street and forms of particular innovative activity in the community. In all this we see a re-engagement of the rational and the public that blurs the distinctions between cognition and emotion, body and mind, sympathy and empathy. It leaves us with an idea of the public as fluid, emergent and as much about the formation of self as the constitution of a wider public. It offers a view of public space that is circulatory and networked, rather than still and symbolic.

RATIONALITY RESTRICTS THE PUBLIC

As Richard Sennett has argued (1974, 2000) Hegel's answer to the question of how an impersonal realm could become a self-constituting public was to link the impersonal with the rational. Simmel (1950) reworks the understanding of the rational from law-making into a social construction. He sees social interaction between strangers in the modern metropolis being characterized by indifference, as urban dwellers hide behind a mask of rationality (neutral exchanges between people in a state of equilibrium amongst strangers: Sennett 2000).

As Sennett puts it:

> It is something that is rooted in the inability of people to live in cities openly as animals. The negative of this Simmelian notion of the mask of rationality is that information is exchanged in this way but communication is lowered, particularly communication that transcends self-interest as well as communication of an emotional sort. The way in which you create an impersonal public realm is by lowering as it were, the amount of information given. What this mask does is restrict the amount that people can know about you. It is reduced to information that can be codified and exchanged like money.
>
> (Sennett 2000: 381)

Urban rationality is a retreat from the unbearable nature of difference articulated through the idea of over-stimulation. Forms of self-identification were reduced to visual signs, such as styles of dress, which could be interpreted instantly without any further mutual exploration (considered in Chapter 3). Visualization and clear categorization were also evident in the broader rationalities of urban planning, involving land use zoning and the creation of a rationally 'efficient' urban system (Boyer 1983; and the subject of Chapter 7). Thus the potential of the city to provide the basis for a self-constituting public, beyond the partialities of associative civic life, were lost. In these conditions urban public space, even where it was generally accessible and a potentially open space of unpredictable encounter, served no broader function in terms of the constitution of a public realm. As Sennett has lamented it is dead public space.

The weight of evidence seems to point to the strangulation of the public realm in the city. Sennett (1974) has portrayed the demise of an impartial public through the death of public space in cities and the replacement of public discussion by the politics of the private realm in the guise of personality politics and the privileging of the intimate. Sorkin (1992) has catalogued the privatization of public space through the malling of America and the transformation of potentially vital public spaces into 'Variations on a Theme Park'. And Mitchell has noted the behavioural regulation of public parks and the protection of individual private 'bubble spaces' as manifestations of the corrosion of the public realm (Mitchell 1995, 2003). The prospects of any kind of self-constituting public establishing itself in the contemporary city seem profoundly bleak. Neighbourhoods are segregated and divided, state regulation and surveillance have reached unprecedented levels, forms of urban indifference seem greater than ever. Yet these very conditions fashion the ways that new public spaces and public realms might be forged. This chapter looks at the lessons of certain historical examples as well as the potential of new forms of associational politics.

RATIONALITY OR PERFORMATIVITY IN PUBLIC?

Habermas's work suggests we cast our gaze to other spaces to reclaim the link between rationality and the public. The sort of instrumental exchanges that characterized Simmel's metropolis and our present-day cities are only one strand of the function of language (and a parasitic one at that). Locution and illocution are, by contrast, communicative competences aimed at mutual understanding that are fundamental to the everyday lifeworld exchanges. Locution is saying something (hello) and illocution is saying something by saying something (saying hello is a form of greeting, and greeting has greater communicative import than simply saying something). The problem as Habermas sees it is that in modern capitalism these communicative realms have become ever more specialized and restricted. Extending the communicative potential of language beyond the domestic arena (constructed as the private) as well as restricted and often conflictual communities of dense association, is a daunting challenge. The possibilities of constituting a wider public beyond these realms is achieved for Habermas in talks about interests that start with statements that reflect difference, the situatedness of lifeworld contexts (in terms of subjective inner states, social norms as well as objective interests). These interests are broadened by the mechanism of an ongoing discussion in which the statements might be challenged at any time in terms of their good reasons or grounds for the argument (communicative rationality). In this conception differences do not disappear but they cease to matter so much in an ongoing dialogue in which talks about interests objectify (and start to transcend) self-interest.

The interaction of people with different situated concerns (of self, local community and worldview) were permitted by 'spaces' in the city (such as the coffee houses of the eighteenth century metropolis) where discussion of the issues of the day, reported in mass circulation newspapers, could be held (Habermas 1989). Sennett makes a different interpretation from Habermas of these nascent public spaces of the city. He starts to separate notions of the public from ideas of the rational. Sennett asks how coffee house customers made their statements credible to each other in circumstances where their social differences were so readily apparent (and read visually). Sennett argues that, rather than subsuming difference in rationality, they dramatized difference by adopting the conventions of theatrical performance, rhetoric, gesture and emphasis, to communicate emotionally and forcefully with each other. This provided a space outside standard modes of communication between different types of people. This theatre of self-dramatization where one could lose oneself in the performance, represented a nascent public realm.

In contrast to Sennett, Habermas's argument of the significance of language in public communication is based on Mead's idea of language as a significant symbol in human evolution and the development of the individual as we have seen in chapters 'on the body' and 'in the community'. The use of shared gestures that build into significant symbols is a process of individual communicators being able to create a response in others that they themselves can recognize and anticipate. Being able to 'take the attitude of the other' also involves seeing oneself as an object. The development of significant symbols, both in human evolution and in individual psychological and emotional growth, involves passing through a stage where gestures are shared between self and a particular other, into a behaviour that is able to take in the anticipated responses of the entire community to which she belongs (Mead's generalized other). This is the social process of rationality, as Mead explains:

> What I have attempted to do is to bring rationality back into a certain kind of conduct, the type of conduct in which the individual puts himself in the attitude of the whole community to which he belongs. This implies that the whole group is involved in some organised activity and that in this organised activity the action of one calls for the action of the other organisms involved. What we term 'reason' arises when one of the organisms takes into its own response the attitude of the other organisms involved. . . . When it does so, it is what we term 'a rational being'.
>
> (Mead 1934: 334)

We saw in Chapter 4 how rationality seems to support the sorts of community-based associational norms of civil society that are partial, rather than any broader conception of a self-constituting, impartial public realm. This tension is reinforced by Habermas's principle of universalization. Although reflective rationality is required to defend subjective, social and objective statements from the challenge to their claims to validity, it assumes that all who can enter into argumentation are capable of coming to the same judgements of the acceptability of norms of action and that 'this is true only if those affected by an action or norm have values, emotions, affections and preferences that are more or less compatible' (Moon 1995: 150). Only then could they 'reach consensus on generalisable maxims' (Habermas 1990: 120, cited in Moon 1995: 150).

Although Sennett positions his dramaturgical understanding of the public realm as an alternative to rational interpretations, I argue his strategy still encounters the problem of the necessity of rationality. Although the dramaturgical device does allow people to communicate emotionally and emphatically with strangers who are different from them, without threat, it only does so because it follows the strict conventions of role-playing of the theatre. The application of these ideas to social life more generally, as seen in the work of Goffman, enables a display of difference but only because these underlying conventions are emphatically rational, and strategically so. These assumptions are called upon tacitly in any encounter between strangers. For instance rationality underlies the common definition of the situation. There is an intimate relationship between strategy and ritual in everyday encounters. There is also a deeper operation of rationality that determines whether the interaction can take place at all. Goffman calls it Felicity's Condition and defines it as 'any arrangement which leads us to judge an individual's acts not to be a manifestation of strangeness' (Goffman 1983: 27). It is the 'rational' presuppositions about presuppositions.

The problem for the ideas of Habermas and Sennett is that their conceptions of the public are hindered by the operation of strategically rational expectations. These conform to a more profound logic of identity, or reason with a capital R. In this sense, escaping self-interest through discussion or dramaturgy does not overcome the 'logic of identity'. The logic of identity confines the idea of the other to an exercise of sympathy (putting oneself in the other's shoes) and symmetry. This idea of sympathy dates back to Adam Smith and his 'impartial spectator' that acts as a check on individual action. The symmetry of perspectives this entails leads to the idea of the generalized other that is apparent in Mead (see Shott 1976). Difference is subsumed in an idea of the other that is a sympathetic projection of the self. This might enable an escape from self-interest but not from ego itself.

Sennett is relying on the performative force of the 'I' and the contribution of individual experience to the discursive community. This is a form of empathy that results from forceful emotional communication. Habermas is relying on the development of cognitive sympathy (in the tradition that goes back to Adam Smith). In both cases individuals are detached from their particular interests, by being tempered by social expectations (in the case of Habermas) or lost in the performative characterization (in the case of Sennett). One is the melioration of difference, the other is its dramatization.

THE PUBLIC AS PARTICULARITY

A radically contrasting set of ideas of the formation of a public rests on an idea of the encounter with the other being based on particularity and asymmetry. Thus Young (1990) famously points to the idea of the particular other that cannot be assimilated through sympathy. The good city epitomizes a space of encounter in which the other cannot be sympathetically grasped – it represents a being together of strangers in conditions of unassimilated otherness. More recently (Young 1997, 2000) posits the idea of asymmetrical reciprocity where the subject accepts the asymmetry of condition between itself and the other and seeks a relationship of reciprocity without trying to put itself in the other's shoes. Young believes this will help preserve difference.

One crucial element of communicative reason, for Young, is greeting. Rather than being treated as the superficial gestures at the edges of more important conversations, greetings are central to the ability to be acknowledged and listened to. The category of greeting thus adds something

important to ideals of inclusive public reason. It is not simply that participants in public discussion should have reasons that others can accept, but
they must also explicitly acknowledge the others whom they aim to persuade
(Young 2000: 62). We saw in Chapter 3 'On the street' how the phatic edges
of communication are vitally important. They are especially important
where encounter involves difference. The structure of communication in a
greeting or cursory encounter is in fact the same as that in more substantive
exchanges of meaning. Greeting itself is a rational convention, a prior
agreement (as in a coordination of expectation) as to what a greeting is. We
have seen how a greeting (such as 'hello') is both a locution (saying something) and an illocution (saying something by saying something). Its illocu-
tionary quality is that it is a commonly agreed symbol of acknowledgement.
Young's insight, and the importance for the idea of transactional rationality,
is that greeting depends on addressing the *particular* individual. Although
based on a common symbol it requires encounter with a particular other in a
concrete situation for its renewal. Using Levinas's distinction between Saying
and Said she argues:

> Levinas's account of Saying is that this acknowledgement cannot come
> in the form of a general appeal to 'all reasonable persons'. It must be
> more particular: I or we must try to persuade you who are in this social
> situation. We must be responsive to you who have this claim on us, listen
> carefully to you, even though we may perceive that our interests conflict
> fundamentally or we may come from different ways of life with little
> mutual understanding.
>
> (Young 2000: 61)

Again following Levinas, Young likens these moments of attentiveness to the
other to being held hostage. As we saw in Chapter 3, there are parallels here
to the interaction rituals that Goffman and others have investigated. Indeed
as Goffman and others have observed, saving the face of the other is a
modern-day and more superficial manifestation of more ancient rituals that
acknowledge the sacredness of the other. Being hostage or being taken
hostage is all too evident in actions that preserve the face of the other when
that other person is using the ritual manipulatively – witness the street
'stemmers' in Goffman's example. These are street salesmen who get people
to commit to buying their goods because purchasing the goods preserves the
face of the seller whose overdone claims would otherwise result in their
embarrassment.

I suggest that greeting is the place in conversation where particularity
meets social convention. In the split second that the person agreeing to buy
the street-sellers' wares to protect their face, they are hostage to the particularities of that person and the situation. In that hostage moment they are
lost to themselves, to their own ego. In that moment there is the undeniability
of the other. But also there is the moment not of loss of self but of ongoing

constitution of self. This is the process of being in public that I think Dewey captures in his notion of rationality encompassing both the spontaneous constitution of the 'I' and the social regulation of the 'me'.

Understandings of the public realm and urban public space have taken the socially regulated me as their starting point. This is a point made by Lofland in her comprehensive research on interaction in public (Lofland 1998). Lofland has argued that the negative view of the potential of urban public space is in part a result of the over-concentration on forms of symbolic interaction. These are the shared symbols that are rationally supported in the relationship between an individual and the expectations of the community to which she belongs. Lofland shows how too much emphasis on the discursive elements of interaction hides a good deal of interaction that goes on in cities. Herbert Blumer (1969) enlarged the distinction between symbolic and non-symbolic interaction: non-symbolic interaction involves immediate and reflex-type reactions; symbolic interaction involves interpretation of the action.

> Non-symbolic interaction takes place when one responds directly to the action of another without interpreting that action; symbolic interaction involves interpreting of the action. Non-symbolic interaction is most readily apparent in reflex responses, as in the case of a boxer who automatically raises his arm to parry a blow. . . . In their association human beings engage plentifully in non-symbolic interaction as they respond immediately and unreflectively to each other's bodily movements, expressions, and tones of voice, but their characteristic mode of interaction is on the symbolic level, as they seek to understand the meaning of each other's action.
>
> (Blumer 1969: 8–9, cited in Lofland 1998: xvi)

In the understanding of the city via symbolic interactionist research it is a concern with symbolic interaction that has predominated. Lofland's point is that this has resulted in an overly pessimistic view of the communicative potential of urban space. By admitting non-symbolic as well as symbolic communication, the ordinary spaces of the city become more promising for the development of a fuller public realm.

Lofland suggests that there is a good deal of cooperation in stranger interaction that goes unremarked. She notes the 'cooperative motility' in acts like civil inattention that are not necessarily asocial, but rather involve a ritual regard for the other person. She captures moments of restrained helpfulness and civility toward diversity. Whilst not being naively optimistic Lofland notes how much of what goes on in the public spaces of the city provide opportunities for a further expansion of communication and regard for the other.

Lofland argues that Blumer downplays the significance of non-symbolic interaction because it is unreflective. She argues that much of what goes on in the urban public realm 'consists of humans responding "immediately" and *apparently* "unreflectively" to others' bodily movements and expressions' (Lofland 1998: xvi, her emphasis). Lofland's research shows how much that goes on at the non-symbolic level involves differing degrees of bodily and cognitive attention. She distinguishes different types of relationships – fleeting (strangers in a street); routinized (customer–seller); quasi-primary (chatting dog owners); intimate secondary (regular bus riders). There are various uses of this ordinary public realm in cities. One is educative – the activity and diversity of a city street can help socialize children (a point made by Jane Jacobs). Lofland argues that public space also provides 'respites and refreshments' – sociable and psychic rewards gained from such things as neighbourhood vitality (where it exists) and the feel of the place. The public realm also operates as a communications centre as well as an arena for the 'practice' of politics

What I am arguing here is that what Blumer did in privileging certain forms of social interaction in cities, Habermas has done for an understanding of the communicative possibilities of cities in the constitution of a public realm. One has narrowed that range of possible interaction seen as significant for the constitution of a truly public space, the other has narrowed the range of communicative rationality that is taken into account in the possible formation of the public realm. Just as Lofland finds more interactive potential in the public realm of the city, so I would argue there are more communicatively rational possibilities in the city. But this involves widening our view of rationality. It also involves an idea of mobile publics in a more fluid conception of urban public space.

The idea of the spaces of the public being opened up by relationships of particularity and asymmetry is expanded upon by Rosalyn Deutsche (1999). Again drawing on the work of Levinas, Deutsche argues that the city could symbolize (but does so all too rarely) the asymmetrical encounter between self and particular, concrete other. The concrete other is the 'I' that is a singularity, the face of the other that suffers and can be named. This particularity of encounter is the moment of undeniable and unrationalizable ethical responsibility: 'responsibility to the Other pre-exists self-consciousness; it by passes rational, calculative thought. I do not grasp the other so as to dominate, but I respond to the face's epiphany as if to a summons that cannot be ignored' (Gardiner 1996: 132). The moment of encounter here is not the measured mutual assessments of interaction rituals but a moment of revelation, an epiphany if you like. This experience of alterity is in the face of infinity rather than any generalized sympathetic totality. This encounter involves non-identification but non-indifference to an inherently unknowable concrete other. This encounter also reveals the incompleteness of the self – it is an escape from ego. Addressing the concreteness of the other is an act of reasonableness rather than rationality (Deutsche 1999).

The city potentially plays a major role in building this ethical public. Its diversity and unpredictability present more occasions for the encounter with the other. The experience of the self as incomplete and unknowable resonates with an infinite and inherently unknowable city. Whereas the rational city, even the deliberatively rational city, was legible, ordered, visualizable, the 'reasonable city' is uncertain, labyrinthine and uncanny. As Deutsche alludes, the subject in process as Kristeva (1991) might call it, is the experience of moving through this city. It is a city of affect, best captured not in maps and plans but in psychoanalytic and literary readings of the urban. Thus Deutsche's call for a 'reasonable urbanism' uses Neil Bartlett's novel *Ready to Catch Him Should He Fall* which she describes as a love story whose main characters include two men and a city. In this story the public statues of London start to speak and move, to shake off their petrified identities as ideal citizens and symbols of power existing in some coherent, democratic totality. At the height of a thunderstorm they rise up to protest against the homophobic violence that denies difference in the city. The device of ekphrasis transforms the objects into subjects of conversation and 'figures the way in which ideas, language and images are reanimated – though practices of use' (Deutsche 1999: 188). With the loss of the ground of certainty, of ego and reason, the grounds of social order also disappear: 'indeterminacy exposes us to others and, with this exposure, democracy is invented' (Deutsche 1999: 184).

'I', 'ME' AND THE PUBLIC

At this juncture I want to gather some fragments from the previous discussion that I think offer a possible zone of engagement between the rational and the reasonable and another way of thinking about the public and the city. First, arguments about reasonableness relate to the long-established notion (from Aristotle onwards) of practical reason. This is a guide to ongoing practical activity in situated contexts, rather than the abstract goal setting of instrumental rationality. Secondly there are the discussions of the role of cognition, the emotions, the unconscious and the body in constructions of the public realm. Third are debates over whether the public realm is (or should be) associational and consensus building or should emerge out of conflict and be agonistic (Arendt 1958; Mouffe 2000). Fourth is the persistent issue of the pervasiveness of instrumental rationality in social life, its role in preventing the establishment of a public beyond self-interest. Fifth is the relationship of humans to objects.

All these themes are live ones in certain strands of philosophical pragmatism and indeed were so in the classical pragmatism of Mead and Dewey. Mead's theory of the evolution of communication is also an idea of the development of mind as a social relation. That was very much in Habermas's thoughts in his formulation of communicative reason. Some critics of Habermas have pointed to the still strongly cognitive function of reason that is required in the judgement of good reasons or grounds in the defence of validity claims. Others suggest the lack of any body presence between speakers (drawing on ideas of intercorporeality from Merleau Ponty – see for example Crossley 2001). As Mead conceived of it, the social development of mind initially takes the form of gestural communication between a child and its mother. Mead has an awareness of the significance of body projection/ schema in the communicative formation of mind, especially the hands as implements for other implements. The body is the repository of habit but it also acts as a sensor to new and potentially problematic situations. Problematic situations, including what we might call the encounter with the particular other, are ones where it is uncertain how the 'I' will respond to the generalized other.

These ideas of 'the public' and 'the other' rest on strong distinctions between the self and the other. The generalizing impulse of sympathy seems to wipe over difference. Alternatively the irreducibility and irresistability of the other subsume the 'I'. Taking the attitude of the generalized other is not necessarily an issue for self-consciousness initially. It is different from Adam Smith and the British Empiricists' idea of sympathy that pervades so many understandings of the other in the formation of an impersonal public. For Mead and Dewey an attitude is a behavioural disposition. It is a potential for action in a certain direction. These tendencies are realized through the body as well as the mind, and on the basis of past contact with particular others, out of which the attitude to the generalized other is built. Part of that realization is the experience of transactional bodies. Realization of the disposition will depend also on the forms and routes of communication and the way that objects bring human agents into relation. Here again there are hints from Mead. Humans and animals act in mediated environments. Mead considers objects as implements in the consummation of the act. Physical objects are made out of reactions to them as implements that exceed the actual consummation of the immediate act. Thus physical objects used as implements in a particular situation can take on wider significance and bring human actors together in different ways.

In his formulation of conditions suitable for the building of public discourse, Habermas seeks to remove instrumental rationality (in the form of perlocution). An understanding of the operations of strategic rationality (from rational choice theory) has the capacity to correct the excessive utopianism, but also assist in the practical thinking on the public realm for both critical theorists of the public and pragmatist philosophers of democracy. Johnson (1991, 1993) and Knight and Johnson (1999) have argued in a series of papers on the possibilities of an understanding of strategic rationality adding an agonistic dimension (along with expressive acts). Indeed strategic action might be an essential part of building up a public. In her objection to the way that Habermas conceives of debate in public where male norms of speech in public exclude women, Nancy Fraser suggest that a flourishing public realm is one in which there are multiple publics or what she calls 'subaltern counter publics' (Fraser 1992). Subaltern counter publics are protected realms for discussion and understanding for those excluded by the conception of a singular public realm. These protected spaces open up the possibility of more emotional and open communication from within. At the same time, from without they are strategic mobilizations using instrumental logic in a wider public realm of engagement with other publics.

Recourse to Mead and Dewey on the ongoing formation of selfhood goes further by denying any significant difference between instrumental and communicative action. Taking the attitude of the other is at once technological–experimental and communicative and meaning enriching (Rosenthal 2002: 215). Ignoring the technical–experimental aspect of taking the perspective of the other has meant that use of tools and instruments, as well as trial and error, have been lost from ideas of building the public. The definition of purposive instrumental rationality has been overly narrowed in contemporary social theory. Notions of the public have been focused too much on the need for acknowledgement in public, in the open spaces of co-presence, rather than in the myriad ways that people are mediated by objects and systems of communication and the potential for publicity in them. Perhaps the self has been lost too, either as an abstract relation or as incomplete respondent to the concrete ethical call of the other. But, as we have learned, for Dewey

> social change is predicated . . . not just on the linguistically mediated 'me' as it makes an appearance in discursive communication but on the instinctive, aesthetic, unpremeditated 'I' that bursts forth on the social scene and makes individual experience valuable to the community as a whole.
>
> (Shalin 1992: 271)

Beyond sympathetic symmetry or asymmetric reciprocity is the need, as Rosenthal, characterizing Dewey, puts it

> [It] involves the ongoing process of learning from and restructuring one's concrete self – in the fullness of its emotions, drives, desires, beliefs, purposes and aspirations – in the process of adjusting to the other. The passions give reason its motivation and focus. Growth of self incorporates an ever more encompassing sympathetic understanding of varied and diverse interests.
>
> (Rosenthal 2002: 216)

This is not a form of sympathy coming from an already constituted monadic cognitive ego, 'for Dewey rationality cannot be isolated from the concrete human being in its entirety' (Rosenthal 2002: 216). That concrete human being is a result of circulatory experiences in social communication. It is a rationality based as much on audacious experimentation as on discursive understanding.

This idea of the constitution of a public also draws on a wider view of communication that has resonance with the city. It sees communication not just as limited to the representational function of language and the way that validity claims can be established and redeemed. It sees communication as constitutive of meaning and involving non-discursive performances as well as linguistic utterances, what Langsdorf calls the fourth dimension (beyond objective, social and subjective knowledge). It is an idea of knowledge as inquiry (following Dewey). It draws on Dewey's understanding of the aesthetic dimension as a potential source of change in the lifeworld. It is based on a processual ontology of pre-linguistic communicative experience which constitutes the objects of communicative action (Langsdorf 2002: 144).

This embedded idea of the constitutive function of communication (of which objects are a situated outcome rather than the reality that language seeks to represent) is more sensitive to the operations of strategic rationality in both a good and bad sense. It sees that particular human interests constitute 'situations' and that they might be in conflict with other interests in other situations. All these diverse convictions are included, rather than attempting, as Habermas does, to exclude 'some convictions (arguments that "lose") in order to establish consensus on others (arguments that "win")' (Langsdorf 2002: 47). This interpretation is aware of the constitution of power and interests within the lifeworld, rather than idealizing it as some innocent space that is colonized by the external and overbearing 'system'.

In his representative take on language Habermas 'denies communicative status to the many ways in which people (and animals) somatically develop their interaction with their environments (e.g. chronemics, haptics, kinesics, proxemics, and vocalics) in ways that cannot be represented, but must be performed by communicating bodies' (Langsdorf 2002: 147). But this wider form of communication 'is particularly efficacious in presenting diverse and unrepresented (indeed, as yet universablizable) goals, including those that Habermas excludes as "strategic" rather than "communicative"' (Langsdorf 2002: 147). This pre-linguistic communicative realm is thus one in which strategic purposes can be forged and changed, rather than relying on the power of argument to focus people on a rational consensus.

These performative actions sit at the edges of communication. Langsdorf is adapting Willard's (1989) theory of argumentation as conversational interaction, a form of communication, rather than being specifications of logic: 'It portrays the practice of argumentation as a way of communicating, grounded in dissensus and marked by the "ethnoconventional" (Willard 1989) negotiation of claims' (Langsdorf 1997: 317). This is highly suited to working within the non-foundational and pluralistic political model that came from the postmodern turn. This broadening of the communicative repertoire that can inform public argumentation is also, claims Langsdorf (2000), more able to realize Habermas's own project than his more narrow focus on representative language. Langsdorf recalls (2000: 22) McCarthy's point that Habermas's entire project, from the critique of scientism to the reconstruction of historical materialism, rests on the possibility of providing an account of communication that is both theoretical and normative, where the 'theory of communicative competence' can 'articulate and ground an expanded conception of rationality' (McCarthy 1978: 272).

This expanded conception of rationality is one that encompasses the spontaneous and unmediated 'I' as well as the communally responsible 'me'. The 'I' was dismissed by Habermas in his use of Mead to develop communicative rationality, as 'too Deweyan' (Rosenthal 2002: 213). Habermas emphasized the 'me', the importance of the 'generalized other' in the consensus model of communicative rationality as well as a representative idea of language.

The 'processual character and ... context dependency of everyday communication' is taken up by the ethnomethodologists, as Habermas himself observes (1984: 124; in Langsdorf 2002: 148). Expanding communication and comprehension in this way involves accepting divergent and novel conceptions, and avoiding their distortion by embedded meanings and associations. This helps expand the discursive parameters.

Dewey saw the local community as the culmination of generalized communication:

> The connections of the ear with vital and outgoing thought and emotion are immensely closer and more varied than those of the eye. Vision is a spectator; hearing is a participator. Publication is partial and the public which results is partially informed and formed until the meanings it purveys pass from mouth to mouth. There is no limit to the liberal expansion and confirmation of limited personal intellectual endowment which may proceed from the flow of social intelligence when that circulates by word of mouth form one another in the communications of the local community. That and only that gives reality to public opinion. We lie, as Emerson said, in the lap of an immense intelligence. But that intelligence is dormant and its communications are broken, inarticulate and faint until it possesses the local community as its medium.
>
> (Dewey 1927: 218–19)

AT THE EDGES OF COMMUNICATION: CREATING A DEMOCRATIC PUBLIC

Communicative rationality emerging from community relations that, in some ways were at the margins of the polity, was given expression by the civic clubs in Progressive era America. In conditions of rapid industrialization and urbanisation in later nineteenth and early twentieth century America, a number of experiments in grassroots democracy sprung up, including university extensions, the People's Institute and the civic clubs.

The various experiments in the constitution of an urban public are traced by Mattson (1998) in his fascinating *Creating a Democratic Public*. By bringing together diverse populations the growing industrial cities of the USA in the early twentieth century provided conditions in which the immigrant could be subsumed and overwhelmed by the ever-growing mass of population. Alternatively the city also afforded the potential for new forms of public encounter and engagements. What the urban democratic experiments demonstrate is the drive to make the public more and more participatory, but through variously informal means of encounter. The City Beautiful movement focused on parks and public buildings as places where citizens could experience a sense of ownership. Aesthetic improvements would engender a sense of pride and that would spill over into city dwellers taking more responsibility for the fate of their environment. As Mattson relates, Tom Johnson, a progressive mayor of Cleveland, tried to remove monopolies over state franchises (such as the streetcar system) and also argued for home rule for cities (to be free from federal interference) as a way of encouraging citizen participation. He also organized tent meetings that were open to the public and discussed political issues. Mattson points to Johnson's realization of the significance of tent meetings as a more informal space of social encounter and engagement.

> In a tent there is a freedom from restraint that is seldom present in the halls. The audience seems to feel that it has been invited there for the purpose of finding out the position of various speakers. There is greater freedom in asking questions too, and this heckling is the most valuable form of political education. Tent meetings can be held in all parts of the city – in short meetings are literally taken to the people.
>
> (Johnson 1913: 82, cited in Mattson 1998: 37)

The meetings were structured such that all speakers had equal speaking time and there were always question and answer sessions after each speech. They drew a broad working class audience and although they were highly structured they were also lively, involving heckling, arguments and searing questions (Mattson 1998).

Other democratic spaces were the university extensions (extra mural lectures) and the People's Forum. The latter started in 1897 in New York when the People's Institute was founded to provide education and debate. In the People's Forum a lecturer spoke to about one thousand people and the lecture was followed by a question and answer session.

> Question-answer periods encouraged clarity from intellectuals and articulateness from questioners – skills necessary for democratic deliberation. . . . Working class immigrants formed the majority of the audience, and the forums produced a lively floor dynamic, with sharp questions interspersed with booing and heckling. Speakers address a variety of important issues . . . municipal ownership, imperialism and American foreign policy (with a majority voting against intervention in the Philippines), tenement housing, socialism and other issues.
>
> (Mattson 1998: 42)

The Forum movement had established over 100 sites in different cities by 1916.

The most significant of the democratic experiments was the civic clubs and social centre movement that started in Rochester (NY) in 1907 and which had spread to 101 cities by 1912. These clubs utilized neighbourhood public schools in the evenings and weekends for leisure activities and political debate. The important distinction from the Forum and the university extensions is that the clubs were constituted and organized by the participants themselves, rather than by experts and professionals. Again the more agonistic form of politics was not excluded.

Mattson argues that the social centres were not nostalgic attempts to recreate a village life and that the activists were striving to create a self-constitutive public in the city that evaded the restrictions of rural life. Putting the efforts of the social centres and civic clubs into a contemporary framework Mattson suggests:

> Certain contemporary social theorists assert that modern ethical theory and political philosophy must be pluralistic and conflicted, not unified. Social centers activists were, of course not aware of contemporary ethical philosophy, but they seemed to have discovered these lessons long before present-day philosophers. From a communitarian tradition in American social thought they drew out the idea that politics grew out of locales where political issues could be made real and dealt with in face-to-face forums. Community consciousness was a necessary basis for a healthy politics. But social centers activists also drew from a liberal tradition that celebrated the principle of public justification and the idea that political equality and rational deliberation – universal principles that transcended local communities – must shape politics.
>
> (Mattson 1998: 76–7)

The social centres helped take republican localism into a broader and more enlightened realm. 'A good citizen thus developed the best side of the modern world's values – autonomy, self-rule and rationality – while transcending the worst aspects of modernity – self-interest and public apathy. By participating in public life, a citizen developed these values gradually' (Mattson 1998: 81). Pateman argues 'As a result of participating in decision making the individual is educated to distinguish between his (sic) own impulses and desires; he learns to be a public as well as private citizen' (Pateman 1970: 25).

One of the early pioneers of the social centre movement was Mary Parker Follett. After reaching the limits of what was then possible for a woman to achieve in academic research, this New England middle class woman turned to urban activism. Mattson traces Follett's reactions to the social centres movement as they unfolded. He argues that based on her experience in the social centre of the Roxbury district of Boston, Follett believed that the democratic citizen should be passionate and open to difference (Follett 1965).

Her practical experiences also informed her contributions to philosophy. As Mattson conveys, she saw the activities in the social centres revealing a group process. The philosophical interpretation was that human beings are social creatures who only become conscious of themselves through their interactions with others. As we have discovered elsewhere through Dewey, this form of pragmatism was a major and novel challenge to western philosophy and its premise of a self-sufficient cogito. According to Mattson, Follett stressed the importance of an idea of intersubjectivity in both the formation of knowledge and in politics. 'A man is ideally free only so far as he is interpermeated by every other human being: he gains his freedom through a perfect and complete relationship because thereby he achieves his whole nature' (Follett 1965, cited in Mattson 1998: 95). Social centres showed intersubjectivity at work (Mattson 1998: 93). Participants became conscious of each other's ideas and ideas were synthesized into political decisions. Follett called this 'interpenetration' a process of 'acting and reacting, a single and identical process which brings out differences and integrates them into a unity ... differences must be integrated, not annihilated, nor absorbed' (Follett 1965, cited in Mattson 1998: 93). As Mattson argues:

> a new synthesis came out of discussion – one that preserved the original insights of different ideas while going beyond them. This process would repeat itself as new discussions took place. Follett believed this to be the essence of the activities found in social centres and what made them a truly creative force in American politics.
>
> (Mattson 1998: 93–4)

The achievement of a higher unity through discussion was Follett's idealistic moment, according to Mattson, but she thought that social centre debates contained elements of idealism and realism (or pragmatism). 'Whereas idealism celebrated unity, realism stressed plurality and difference' (Mattson 1998: 94). Follett also addressed her interpretation to social theorists by contrasting group processes and 'interpenetration' with the law of the crowd. Social thinkers such as Le Bon (1966) and Park (1972) had seen social gatherings in the urban setting as potentially regressive and irrational, showing a mob mentality.

In contrast to the irrationality of crowd behaviour, the group process Follett identified in the social centres suggested a thoughtful, rational and deliberative form of urban social organization. It has to be said that the concern about the crowd might also have registered the fears of the powerful elites of the potentially revolutionary impact of mass mobilizations. The social centre movement itself has been widely criticized for being a form of social control, in which the potential rebelliousness of the urban working class was dissipated through the strictures of responsible debate in the civic clubs.

There are other messages to take from the social centre movement. They seemed to permit a synthesis and transformation of ideas through difference. They were neither simply communitarian nor liberal but often more radical. From the perspective of the central concerns of this book they also suggest a form of emergent rationality. Contributions were not carefully specified. They were not necessarily reliant on substantive validity claims that won out over other validity claims in a way that Habermas would wish. Rather deliberation was transformative, and in two senses. The first was that the dynamics of debate often led to outcomes that were unpredictable and beyond the discursive positions of the participants. The second sense of transformation was the more gradual one of the acquisition of the skills of deliberation through the practice of debate and through an ongoing programme of education, although these were aspirations *from* participation, rather than qualifications *for* participation. At their best the social centres and civic clubs were educative, open as well as discursive and performative. In the latter sense passion, volubility, and strong expression through gestures were an integral part of the discussion. These discussions were not an exercise in community solidarity. Follett was not a communitarian and indeed believed that true democratic discussion should go beyond community.

So the conditions of discursive plurality in the social centres are derived from the presence of social diversity in the urban neighbourhood. Yet for Follett pluralism itself was too focused on the group, rather than accepting a broader public beyond the group. In a way cannily prescient of Iris Young she wanted to avoid the egoism of the group (Follett 1965). Her call for openness and the diversity of the neighbourhood against contentment of sameness that leads to a meagre personality too resonates with the early work of Richard Sennett. In *The Uses of Disorder* Sennett (1970) argues that retreat into homogenous community stunts psychological development, the healthy, mature version of which is founded on the encounter with difference in the unpredictability of the city. It also relates to the hybridity of the city.

Follett recommended a form of federalism but based on neighbourhoods rather than states in a succession up from the neighbourhood to the city that would federate into states and states into the nation. The legitimacy of the system was based on the self-constituting publics of the neighbourhood associations (Mattson 1998: 97, 98). The neighbourhood was also the core unit for her vision of political education.

The intersubjective nature of deliberation also includes moments of pure self-expression. Self-expression should not be a letting off of steam but can be 'used constructively for the good of society' (Follett 1965, cited in Mattson 1998: 100). It is the 'I' relating to the 'me' and suggests how deeper communicative rationality was emerging in the social clubs (although it was not to last).

What Follett describes in the urban social clubs articulates Dewey's idea of sympathy being a form of action and a process of growth: 'Growth of self incorporates an ever more sympathetic understanding of varied and diverse interests' (Rosenthal 2002: 216). This is not a form of sympathy coming form an already constituted monadic cognitive ego: 'for Dewey rationality cannot be isolated from the concrete human being in its entirety' (Rosenthal 2002: 216). That concrete human being is the result of circulatory experience in social communication, in its broadest sense through bodies, gestures, happenings and discourse.

CIRCULATION THROUGH NETWORKS AND MULTIPLE PUBLICS

Difference theories stress the significance of culture and linguistic styles in differentiating or preventing participants from entering into the kind of discussion that Habermas envisages. Nancy Fraser (1992) has shown how differences of gender, class, race and ethnicity lead to practical differences in styles of communication, and these are shot through with power relations that mean that even if you manage to get yourself heard, you might not be listened to. Fraser points to Pierre Bourdieu's work on distinction to suggest how these differences might be worked upon in separate spheres such that the disempowered may gain some voice. Bourdieu's (1984, 1991) argument is that social divisions shape linguistic styles so that language cannot get behind this to forge some kind of consensus. Certain styles are based on class power and therefore contain 'linguistic capital' that dominates the debate. Speakers 'lacking the legitimate competence are de facto excluded from the social domains in which this competence is required, or are condemned to silence' (Bourdieu 1991: 55).

These differences can extend beyond language to the more performative aspects of communication. They include the symbolic dimensions of communication and dramaturgical, artistic and expressive communication (Fraser 1992). These elements of expressive communication can both help constitute counterpublics but also cross over different public realms.

In their investigation of the nature of the public realm Emirbayer and Sheller (1999) cite Paul Gilroy's description of self-identified 'black publics' where 'dramaturgy, enunciation, and gesture – the pre- and anti-discursive constituents of black metacommunication' are important parts of black counterdiscourses (Gilroy 1993: 75, quoted in Emirbayer and Sheller 1999: 155). Studies of performative communication also point to the 'dispersal of the agon [conflictual relations]', with 'multiple locations or moments of public debate' (Honig 1992, cited in Emirbayer and Sheller 1999: 155).

I think the point that Gilroy makes is particularly germane for the discussion of the edges of communication. He argues how the history of slavery meant that the black voice was literally silenced and that expression and communication took other forms, through the blues, through jazz music, though literature. Gilroy suggests these are the kinesics (communicative body movements) of post-slave populations. Antiphony (call and response) is the modality at work. Antiphony is a form of encounter:

The intense and often bitter dialogues which make the black arts movement offer a small reminder that there is a democratic, communitarian moment enshrined in the practice of antiphony which symbolises and anticipates (but does not guarantee) new, non-dominating social relationships. Lines between self and other are blurred and special forms of pleasure are created as a result of the meetings and conversations that are established between one fractured, incomplete, and unfinished racial self and others.

(Gilroy 1993: 79)

A widened idea of communication and the idea that the public is in process, in the same way that self is emergent, is I think critical to any re-evaluation of the public realm. It requires not the rational deliberation of already formed egos but at once the loss and recovery of self and others. The rationality at work here is held in tension by the impulses of the 'I' and the expectations of the 'me'. This is strongly analogous to the understanding of the inner dynamics of jazz music.

There is a cruel contradiction implicit in the art form itself. For true jazz is an art of individual assertion within and against the group. Each true jazz moment . . . springs from a contest in which the artist challenges all the rest; each solo flight, or improvisation, represents (like the canvasses of a painter) a definition of his [sic] identity: as individual, as member of the collectivity and as a link in the chain of tradition. Thus because jazz finds its very life in the improvisation upon traditional materials, the jazz man must lose his identity even as he finds it . . .

(Ellison 1964: 234, cited in Gilroy 1993: 79)

Jazz reason emerges in the flow of the music and the improvisations that are at once permitted by and rework the structure. Gilroy analyses how certain forms of black music have circulated and transformed in a form of diaspora from the tour of the jubilee singers in the 1870s through to modern rap.

The networked and circulatory nature of communication and the possibilities of forming multiple public realms have been explored by Emirbayer and Sheller (1999). They point to Harrison White's (1992) network approach that identifies publics occurring in liminal spaces and switching locations between more established network domains. I have elsewhere used White's approach to 'personhood' to make the point that identity can emerge in the circulation through time-space power networks (see Bridge 1997b). From their perspective Emirbayer and Sheller argue that publics can be interstitial, networked and multiple. They compose a helpful typology of publics that reveals some of the key dynamics. From the performative model of publicity they draw out power distinctions between dominant and counter-hegemonic publics, which deny Habermas's ideal speech situations, but describe the differences that must be worked across to achieve any procedurally valid agreements (although one could argue that Habermas understands the ideal speech situation to be the outcome of crossing these differences).

The second continuum of categorization is time-space distantiation. This refers to the reach of publics in time and space, from face-to-face communication, through to locally mediated interactions to far-reaching and long-lasting publics based on print or symbolic communication. The third dimension is the effects publics have within civil society but also on economics and politics. Civil, economic and political publics all 'feature some combination or other of strategic and communicative action (in Habermas's terms)' (Emirbayer and Sheller 1999: fn 76, 160). They propose a network analysis of the transactions in these multiple, condensed and distantiated, enclosed and interconnected, emergent publics. They suggest this analysis across the spheres of socio-structural, cultural and social-psychological contexts of action.

There are some interesting elements of Emirbayer and Sheller's prospectus for research on publics that are significant for the present discussion. In the cultural sphere they note how certain discursive practices, both dominant and counter-hegemonic, can structure the public. Mustafa Emirbayer's own research on civil publics seeking school reform in the antebellum and Progressive eras 'were structured in part by discourses of "virtue" and "corruption" in which opponents of educational change became depicted as partisan and self-interested, standing "outside" the sacred center of American moral culture' (Emirbayer 1992, cited in Emirbayer and Sheller 1999: 171). The social-psychological sphere is fundamentally transpersonal, transactional and adapting Randall Collins (1981) involves the emotional economy of such transactions that take place in situations (see Chapter 2 on transactional bodies). These elements help to theorize the 'non-rational sides of publicity: that is, its group dynamics, leader-follower ambivalences, unconscious re-enactments of a "family romance', and so forth' (Emirbayer and Sheller 1999: 175).

John Dewey thought ideally of the public realm being democracy as a way of life, the idea of community life itself. He observed in the early twentieth century how mass media often distorted and discouraged full engagement in democracy on behalf of the public. Social intelligence that would realize its expression through neighbourhood and community engagement was running idle. We might argue, along with Habermas, that the dominating elements of the system are now immensely more oppressive than they were in Dewey's day. There are dominating discourses within the public realm and disciplinary surveillance of public space. Yet there are modes of communication in more fluid spaces of the city in which public building activities go unremarked. They are more tentative and temporary communicative impulses, often at the margins, but ones that need to be recognized and understood as part of any more socially encompassing idea of the public realm.

6 At work and home in the urban economy

RATIONALITY, CULTURE AND THE URBAN ECONOMY

Economic rationality rules the city. From the political left or the right, as a Marxist or a liberal, it is the imperatives of instrumental economic rationality that have the greatest impact on the built form, physical and social divisions and the pace and pulse of the city. Yet the character of this rationality is strangely abstract, separate from the people and the cultures it makes wealthy or impoverishes. In the neoclassical model of urban form people are reduced to consumers who bid for land in the same way as firms do. In Marxist urban studies it is the logic of capital that takes precedence; people and cultures take their form from economic imperatives.

There are possibilities of resistance to the overall economic logic. As we have seen in previous chapters resistance has been attributed to seemingly non-rational spaces (such as of the body) or sites of alternative rationalities, such as the communicative rationality that Habermas proposes. Habermas's alternative rests on a sharp dichotomy between strategic and communicative rationality and their respective spheres of operation in 'the system' and 'the lifeworld'. One is where economic rationality operates and the other has the vestiges of the rationality that underpins everyday language use to establish understanding and is not concerned with instrumental gain. In this perspective 'culture' has been seen as being separate from the economy, home to more traditional and important shared values and loyalties beyond the narrow acquisitiveness of economic rationality. Understandings of urban change that privilege such aspects of city life as lifestyle and culture have tended to be distinct from economic readings.

There is a growing trend to read the economic and the cultural together. An example is where capital is seen to absorb culture, such as in Sharon Zukin's compelling account of the transition from garment factories to artist lofts to gentrified 'loft living' in the Lower East Side of Manhattan (Zukin 1982) and her broader reading of the role of culture in urban economies (Zukin 1995). Here the relationship is one of appropriation rather than interpenetration: the transition from one logic to another. Culture is used in the advertising and place marketing of cities (Kearns and Philo 1993). In this case aspects of culture are reduced to relational assets in the competition between cities for inward investment. Culture and lifestyle are collapsed onto the instrumental rationality of inter-urban competition. There are other examples of this relationship in the tourist city (Judd and Fainstein 1999).

This chapter suggests a reading of the urban economy that mutually implicates traditional economic factors with lifestyle and culture. It emphasizes taste and discernment, rather than preferences and choice, along with habit and cultural particularity, in the stabilization of economic institutions. It suggests that economic divisions are not just the result of impersonal forces such as capital accumulation or globalization, but are deeply implicated in ongoing ways of life. It suggests how the 'hard facts' of economic rationality (looked at from either a neoclassical or Marxist perspective) are aligned with the semiotics of communication. It looks to a critical symbolic economics to reveal the phantasmagoric underpinnings of socio-economic life that has its strongest manifestation in the urban environment. It suggests how the real and virtual spaces of the postmodern city (Soja 1996) relate to earlier readings of the phantasmagoria of the urban economy (Benjamin 1979). It suggests that in this hyperreal space there are relative stabilities, economic institutions that are socially constructed invariants in a 'cumulative sequence of habituation' (Veblen 1961). This approach stresses the significance of a situated economics and the continued relevance of pragmatic philosophy in decoding the contemporary urban economy.

These moves to understanding the importance of culture and institutions in economic processes have of course been presaged by the new institutional economics (Williamson 1985; Hodgeson *et al.* 1993) and the new socio-economics more generally (Smelser and Swedberg 1994). But I believe they are also inherent in the changing understanding of instrumental rationality itself. Neoclassical economics rests strongly on the assumption of expected utility maximization in an environment that is parametric (i.e. the generalized demand that the market represents can be treated as a given) and usually results in a unique equilibrium. But this is Robinson Crusoe economics: individuals can act without having to take other people into account. Where rationality is acting in a strategic environment (i.e. having to anticipate the choices of other thinking minds, who are also trying to anticipate . . . etc.), coordination becomes more difficult (even if desired). Outcomes often result in multiple equilibria and even where those equilibria are optimal (everyone's best response given what they each expect the others to do – the famous Nash equilibrium), they may not be unique. Where there are several Nash equilibria, something other than strict rationality has to decide – usually precedent, or convention or 'culture'. What the understanding of strategic rationality in game theory does, I argue, is to point to the inherently social nature of rationality. Even though it is based on assumptions of individual cognition over expected utility, the process of meeting other minds and trying to anticipate their moves knowing that they are doing the same, results in technical terms in different types of equilibria from the standard neoclassical model, but in more qualitative terms, the way that choices are made gives rationality an emergent quality. It comes from between the participants in interaction and is not the property of any one of them (see Bridge 1997a, 2000). What this suggests is that there is much closer ground between the idea of the supposedly invariant instrumental rationality from economics and the more communicative, contextual and pragmatic understanding of rationality being pursued here.

This reading of rationality is explored in the urban economy primarily through an example of socio-economic transition that has been emblematic of theoretical disputes and empirical instances of economic change in the city: the case of gentrification. The literature on gentrification has demonstrated the theoretical differences over Marxist versus liberal, capital versus culture, production versus consumption in explanations of economy of the city. What I suggest is that these forces are mutually constitutive and in a way that implicates strategic rationality, as well as communicative rationality in culture, and communicative rationality as well as strategic rationality in economic activity. This has both positive and pessimistic outcomes in thinking about resistance to capitalist imperatives.

GENTRIFICATION: A CASE OF COMPETING ECONOMIC RATIONALITIES

Understandings of gentrification have been divided between, on the one hand, theoretical approaches that suggest the importance of culture or lifestyle and the changing preferences of the new middle class as the primary cause of the process (Ley 1996). Alternatively Marxist analyses of gentrification stress the importance of flows of capital and relative values of land in the central and outer city resulting in a 'rent gap' (Smith 1979, 1996). In the cultural explanations consumption is emphasized and purely economic explanations do not figure strongly. The emphasis is on the production of gentrifiers. In the Marxist explanations economic forces are seen as primarily responsible for the production of potentially gentrifiable dwellings or land. Is gentrification a case of developer/real estate agents' economic rationality in response to broader opportunities presented by the circuits of capital? Or is it an outcome of changing preferences in the urban housing market for a section of the new middle class? If it is the latter to what extent can these preferences be said to be purely economic and instrumental or are they based on less tangible, aesthetic tastes? Gentrification is a prominent case of the debates over the nature and effects of economic rationality in the city.

Gentrification also poses a conundrum for the traditional neoclassical economic models. Alonso's land use model assumes that wealthier residents will live furthest from the centre of the city (Alonso 1964). Commuting costs are least important for this group (from within the band of residential consumers in the model). What is assumed is that wealthy residents will prefer the larger housing space available further from the city centre to the commuting costs (including time) to get to their jobs in the city. These assumptions are also behind the filter-down model of change in the housing market whereby with industrial growth middle class residents move out of the city and the housing they leave filters down through the social classes (Hoyt 1939). Eventually through a process of sub-division of property and change of ownership central city housing is dominated by bedsits for poorer renters. Gentrification seems to contradict the assumptions of filter-down of housing. The move downtown by certain members of the middle class and their willingness to 'do up' filtered-down housing poses a contradiction, especially since the majority of middle class residents conform to the assumptions of the models by continuing to suburbanize.

One explanation consistent with the neoclassical canon was that gentrifiers are the leading edge of a wider metropolitan turnaround (Laska and Spain 1980). Parallel to this was the explanation that rising commuting costs and traffic congestion were beginning to bear down on the preferences of wealthy residents, some of whom were adjusting their residential decision making as a result. Demographic factors, including the emergence of young urban professionals (Yuppies) (Blum 1983; Short 1989) and gay gentrification (Lauria and Knopp 1985) have also been identified. Another set of related explanations points to the increasing attractiveness of the postindustrial city as residential location (Ley 1996). However, the effects of the shift to postindustrialism are very uneven: de-industrialization has rendered the centre of many cities just as, if not more, undesirable as they were when noxious industries were located there. That qualification brings in questions of the placement of the city in the urban hierarchy in an increasingly global economy. Wide scale gentrification is a classic marker of the global city (Sassen 2000); there is also growing evidence of more diffuse types of gentrification in provincial cities (Bridge 2003; Dutton 2004).

The shift to a postindustrial city leads to economic explanations for gentrification that hinge on the increased numbers of new middle class occupations. The rise in numbers of the new middle class and their ability to outbid most other residents in the urban residential land market is a strong explanation of gentrification. Indeed Hamnett argues that the growth of this new class is the primary explanation of the process. This is a strongly materialist explanation Hamnett avers (1991, 1994) because it is grounded in economic change rather than the individual preferences of the gentrifiers. Where it differs from Smith's materialist 'rent gap' explanation (see below) is in proposing that it is occupational shifts in the urban labour market, rather than the dynamics of the urban land and property markets, that is the key driver of the process. Hamnett points to many cities (for example Washington DC) where the more professionalized workforce have continued to suburb-anize and where the rent gap has not been closed.

One set of arguments denies any significance to consumer preferences in the explanation for gentrification. Neil Smith argues that gentrification is a 'movement of capital, not people' (Smith 1979). Smith's Marxist analysis looks at the urban land market and the relativities of investment. After a period of de-industrialization and suburbanization in the 1970s and '80s the central city land market of many western cities had become depressed. Central land was often vacant and derelict. Accepting the neoclassical assumptions about the accessibility of central city land making that land potentially valuable, Smith pointed to the rent gap between the actual value of central city land relative to its potential value under the highest and best commercial uses. Given the rationality of investor action described above there came a point where the rent gap was so large that central land once again became an attractive investment prospect. Gentrification was a way of closing the rent gap. Thus the gentrification process was a developer-led movement of capital not people. Aside from questions about whether these rational decisions were net of risk that such a large rent gap would surely imply, this analysis leaves open the question of how certain sections of the middle class might be persuaded to come and live in the central city when their established preference was to move to the suburbs. As well as the production of gentrifiable housing there is the question of the production of gentrifiers.

Sharon Zukin's (1982, 1995) work takes account both of the rational agencies at work in the urban land market as well as cultural and social dynamics that might account for the production of gentrifiers. What Zukin argues is that cultural changes, however subtle, become potential sites for commodification and renewed rounds of capital accumulation. In the 1970s in the Lower East Side of Manhattan de-industrialization meant that the lofts that had housed small factories, especially in the garment industries, fell vacant. The lofts were then occupied by artists, who either squatted or paid low rents. As the reputation of the district as an artistic centre grew, real estate agents became interested, seeing an opportunity to market the area as a safe bohemia to wealthier Manhattan residents. Developers started to buy up the properties and convert them into luxury apartments. In the process the artists were displaced and only the most exclusive galleries remained in the area. In this process capital captures culture. The economic rationalities of poor artists that brought them to the Lower East Side gets displaced by a more powerful rationality of investment in the urban land market as a whole, via the artist's artistic practices.

Zukin's is a case where there was reasonably rapid investor-developer involvement once the TriBeCa neighbourhood had become established as an artists' enclave. The risks were borne by the developers as lofts were converted into exclusive apartments. The ground was thus prepared from which to encourage the gentrifiers to settle. This sort of developer-led gentrification is also typical of many of the waterfront, docklands-type developments such as Sydney's Darling harbour, Battery Park City, New York and London Docklands. What also has to be explained is the more piecemeal activity of small builder-developers and owner-occupiers gentrifying the properties of some of the more historic central city neighbourhoods of many western cities. This is the original form of gentrification noted by Ruth Glass in London in the early 1960s (Glass 1964) and continues to be a powerful force shaping the social geography of cities today. What motivates household gentrifiers to move into central city neighbourhoods, especially when the economic assumptions are that they prefer outer city space to the higher density accommodation in the central city?

As we have noted, some of the explanation might rest with traditional economic factors such as increased commuting costs (time and money) with long working hours of typical gentrifier occupations making commuting a particular burden. Nevertheless given the rate of rising real incomes in these sectors in is unlikely that commuting costs would be a deciding factor. In the bid-rent city it is more likely to be choice rather than constraint that determines the residential location of the new middle class. Some researchers have looked therefore to a broader set of lifestyle and value considerations.

David Ley (1980, 1986, 1996) has been crucial in suggesting how a new set of pro-urban preferences might get established. Present-day gentrifiers, brought up in the 1960s when the peace movement and hippy movement were in their heyday, have inherited a different set of values of nonconformity. Many of them received their higher education on campuses in the centre of large cities and were involved socially and/or politically in those places. Their political and social orientations also fed into their habits as consumers. The mass production/mass consumption regime of their parents' generation, including mass consumption in the mass-produced suburbs, was resisted. Historic goods (such as second-hand items) and historic housing was valued over the developer-led suburbs. Access to the social life and entertainments of the central city became more important. This was especially the case with the rising numbers of professional women workers and the delayed fertility that resulted. The combination of these factors has resulted, according to Ley, in a strong pro-urban aesthetic amongst a section of the highly educated middle class (see also Mills 1988).

Distinctive aesthetic tastes have been identified as one of the defining features of this new middle class. Some researchers go so far as to argue for a distinct gentrification habitus (Podmore 1998). Podmore describes how the idea of loft living became an internationally marketable commodity because it called upon a set of dispositions of distinctive, residence, lifestyle and consumption. There are arguments that link this to class dynamics in the mix of economic and cultural capital and the way they are deployed in the landscapes of gentrification (Butler 1997; Bridge 1994, 1995, 2001a,b).

Using the case of gentrification to think about how economic rationality shapes the city raises questions about the degree to which that rationality is broadly or narrowly defined and whether it is thought to be a conscious or unconscious process. Neil Smith's Marxist analysis put the explanation squarely in the unconscious logic of capital accumulation and the relativities of value in the urban land market. This explanation is not done and dusted in the sense that investor-developers acting as agents of capital must make judgement calls about whether risky investments in inner urban land will pay off, and in what way. If investment for gentrification is seen as the route there must be a series of conscious calculations that take in ideas of the social receptivity of gentrification in the city in question. David Ley offers a conscious preference-based account but one that takes in a whole series of liberal values and ideas of the liveable city. What Ley's account lacks is a mechanism to suggest why, given these gentrifier predispositions, gentrification happens in some cities rather than others. Hamnett looks to the professionalization of the labour market in the postindustrial city as the explanation (narrow economic, unconscious explanation) but cannot deny the role of education and culture in explaining why some new middle class residents suburbanize. Sharon Zukin demonstrates how narrow economic rationalities (artists' need for affordable and suitable accommodation in Manhattan) establishes a set of aesthetic practices that are subsequently commodified by the narrow economic rationalities of investor-developers in the place marketing of artistic Lower East Side to wealthier professional residents.

While Zukin suggests the sequential ordering of broader cultural practices subsequently commodified by narrow economic rationality, what I am arguing for here is the persistent interpenetration of conscious/unconscious, broader and narrower economic rationality. Economic processes and practices involve both strategic rationality and communicative rationality – what elsewhere in the book I have called 'transactional rationality'. In economic terms the basic unit of analysis is the transaction, in this case the communicative elements of gentrification that signal good taste and class status. Gentrification is expressive in the sense that it consciously registers the aesthetic and lifestyle preferences of this section (or perhaps cohort) of the new middle class. It involves rationality in two senses. The first is that it involves a set of economic and instrumental calculations by both gentrifiers and other agents in the gentrification market. Secondly, and relatedly, it engages rationality as a mechanism that holds in place a relatively stable set of expectations that people can have of each other. Transactional rationality explains a new set of signalled dispositions and preferences that have shaped and continue to shape the city and how they rapidly construct a set of aesthetic, as well as economic, borders that draw new social divisions in the city.

TRANSACTIONAL RATIONALITY AND THE SITUATED ECONOMY

In an important contribution to the debate Michael Jager (1986) suggested how the gentrification aesthetic of Victoriana in Melbourne was reminiscent of Veblen's arguments about the significance of leisure and conspicuous consumption in maintaining social status. As Jager puts it:

> For Veblen's leisure class, servants had a dual function; they had to work and perform, and they also had to signify their master's standing. Gentrified housing follows a similar social logic. On the one hand, housing has to confer social status, meaning and prestige, but on the other it has to obey the social ethic of production: it has to function economically. This unites the performance ethic and the signifying function.
>
> (Jager 1986: 79)

Gentrifiers like the bourgeois in Veblen's time are caught between emulation of upper class consumption habits and avoidance of the economic necessity of work. In gentrification this takes the form of a selective and 'tasteful' restoration of property and re-imagineering of history to distinguish themselves from the working class.

What I wish to develop here is the fact that Veblen's argument about conspicuous consumption was in fact a critique of the utilitarian assumptions of neoclassical economics and more specifically utilitarian rationality and the hedonistic assumptions on which it was based. As Veblen argues:

> The hedonistic conception of man is that of a lightening calculator of pleasures and pains, who oscillates like a homogenous globule of desire of happiness under the impulse of stimuli that shift him about the area, but leave him intact. He has neither antecedent nor consequent. . . . Spiritually the hedonistic man is not a prime mover. He is not the set of a process of living, except in the sense that he is subject to a series of permutations enforced upon him by circumstances external and alien to him.
>
> (Veblen 1961: 73–4).

The levels of consumption observed by Veblen were excessive, given utilitarian assumptions. Emulation of the leisure class was costing the bourgeois dear, way beyond what would have been expected of instrumentally rational consumers. The reason for the excessive consumption is the need to signal to others the social standing of the possessor.

For Veblen the use of certain forms of consumption to communicate social standing is not one that is formed in the interaction between atomistic individuals but is rather an evolving whole 'a cumulative sequence of habituation' (Veblen 1961: 241) in an evolving social culture. The basis of this evolutionary model is that of natural evolution and an understanding of economic activity based on an organic model. This contrasts with the mechanistic assumptions behind neoclassical economics. Rather than the means–ends calculations of atomized individuals adding up to a generalized demand (as in the bid rent model) or movements of capital simply bearing down on atomized and alienated individuals, in this approach means and ends are an emergent and changing property of the transactions in socio-cultural complexes that form the basic institutions or 'socially constructed invariant' (Mirowksi 1988) of the economic system. This institutional approach emphasizes ongoing processes rather than abstract universals, habit and routinized behaviours rather than rational actions, and rather than the search for equilibrium states deriving from classical mechanics it sees the economic system as ever-evolving, driven by habit but open to innovation.

Socio-cultural complexes are the key institutions of economic activity. Economic activity is the result of shared dispositions and transactions rather that atomistic individual choices over the consumption of different goods. Rather than being driven by rational calculation, economic activity can be the result of deep desires for status and recognition. Conspicuous consumption reveals that consumption is tied to inequality, because status is premised on inequalities between people. This militates against any egalitarian aspirations in a growing economy. The consumption of inequality also, for Veblen, suggests an inherent instability of the economic system. In terms of the argument I am pursuing these status struggles suggest how the resources of communicative rationality can be used competitively and add to social divisions – over taste and body comportment as well as the more obvious material inequalities.

If economic activity is largely a result of cumulative habit and disposition then the work of Pierre Bourdieu (1977, 1984) and his ideas of distinction and an economy of practices is highly significant. Whilst Bourdieu's idea of the logic of practice captures the economic in its broadest sense (discussed below), his idea of how these processes operate is decidedly non-cognitive. Gentrification can be seen as a collection of class strategies involving varying deployments of cultural and economic capital in time and space (Butler and Robson 2001; Bridge 2001a,b). Economic and cultural capital can be exchanged for one another in various ways and may at various times operate to reinforce class power, or alternatively to act as contradictory forces within the class dynamic. For instance Jager argues that 'the gentrifier is caught between a former gentry ethic of social representation being an end in itself, and a more traditional petty bourgeois ethic of economic valorisation' (Jager 1986: 83). Early gentrifiers lacked the economic capital to achieve distinction and unfettered social display so they substituted cultural capital by making an historic aesthetic out of downtown housing. This aesthetic reinterprets working class space converted into displays of good taste as an act of distinction from their working class neighbours and from the mass production/ mass consumption suburbs. It operates on the social construction of an idea of an authentic neighbourhood.

According to Bourdieu's logic the mechanisms responsible for habitus are largely unconscious. Habitus is an array of inherited dispositions that condition bodily movement, tastes and judgements, according to class position (Bourdieu 1984). It is both a framework for the categorization of things by taste and the classification of those things. It is individually embodied and socially shared within class boundaries. The motive force that reproduces habitus is the drive to maintain distinction in struggles over social space. The key point here is that for Bourdieu the directives of economic rationality are carried out unconsciously and in a more general economy of practices than economic activity traditionally conceived.

Gentrification represents a new field of possibilities in social space that is opened up by changing economic conditions (including the devalorization of inner urban land and the investment opportunities that affords, along with the growth of the service class of professional workers). The point is whether these changes are new and distinct, defining a new middle class, or whether they arise from prior dispositions of class habitus. If the latter is true then gentrification is not a distinct new habitus and the practices it encompasses are a variant of a well-established middle class habitus.

Honneth (1986) has argued that Bourdieu's habitus in fact rests on instrumental criteria of utility maximization. Different groups seek to maximize their utility in the narrow field of the economy as well as in the more general economy of practices that includes symbolic forms. Groups seek to maximize their possession of rare symbolic goods for the least economic expenditure. But rather than the traditional idea of individual utility maximizers, Bourdieu argues for a positionally based utility calculus. This social utility calculus is 'manifest in their collective perceptual and evaluative schema on an unconscious level' (Honneth 1986: 57). As Honneth argues this 'allows Bourdieu to claim that, even if they subjectively orient their actions in other ways, social subjects act from the economic view of utility which has been deposited in the modes of orientation, classificatory schemes and dispositions binding to their groups' (Honneth 1983: 57). So subjective intentions can diverge from habits but habits make sure that group distinction is maintained:

> . . . the forms of life and taste dispositions which different groups, at any given time, pass on through cultural socialisation have a purely instrumental function. They so adapt individual group members to their specific class situation that these individuals, as a result of their valuations and judgements of taste, carry out the appropriate strategic actions aimed at the improvement of their social position.
>
> (Honneth 1986: 63).

The positionally based utility calculus is Bourdieu's idea of what I have been calling the socio-cultural economic institution. This institution is responsible for the 'collective perceptual and evaluative schema on an unconscious level' (Honneth: 1986: 57). This positional utility theory takes into account the status and symbolic elements of consumption and how, in this case, a gentrification habitus can build up and come to represent a major economic force shaping the contemporary city and the planning and political imagination.

The problem with an approach involving group assessment of the objective probabilities that govern their lives is that it does not obviously explain how gentrification got started. Before gentrification prior dispositions over objective probabilities would have ruled out living in run-down working class housing in the inner city as all the neoclassical economic models were happy to assume. The explanation for this encompasses both production and consumption, capital and culture. First it was necessary to have relatively devalued central urban land and dwellings, although the significance of this could be accounted for by both Marxist and neoclassical models. In the Marxist explanation the rent gap provided by devalorized central city land represents an investment opportunity. In the bid-rent model the same devalorized land provides an opportunity for residential bidders to bid for central city land. The problem with the bid-rent model is that it privileges the preferences of the wealthier residential bidders for more housing space in the suburbs and this seems to be reversed in gentrification: so even if they could, why would they bid for central city residences? This rigidity of the neoclassical model of urban land use to account for socio-economic change is symptomatic of a broader difficulty in this regard in neoclassical economics as a whole and again one of the shortcomings that institutional economics sought to address (Barnes 1996).

What may start to account for change in choices and action in the case of gentrification is the diminishing distinctiveness of middle class consumption over the period that gentrification was developing (from the mid-1960s onwards). From a Veblenian perspective the move to the suburbs after the Second World War could be seen as an emulation of a gentry lifestyle in terms of the search for housing space, but also the distinctiveness of a consumption environment, that is village life, where the status of individuals was common knowledge. As Veblen had argued, in the city other communicative devices (such as fashion and conspicuous consumption) had to operate as markers of social status in an environment where most people were strangers to each other (see also Simmel 1950; Sennett 1990). What the mass production of the suburbs meant was an increasing lack of distinctiveness of housing form as a marker of middle class distinction. But it was not just the production of housing that started to deplete the middle class distinction. As Redfern (1997a,b) has pointed out the increasing availability and affordability of domestic technology (such as white goods) meant that middle class families had more money to devote to the aesthetic and performative elements of their housing. This for Redfern is the production factor that creates the opportunity for sweat-equity or commissioned renovation of older properties that leads to gentrification. It is older housing in the inner city that affords the greatest opportunity for renovation and restructuring to make this social display, especially compared with new-build housing in the suburbs. It is also possible that the increased availability of domestic technology flattened status distinctions based on consumer durables that added to the lack of status distinctiveness of suburban life.

What we have been discussing are the potential push factors away from the suburbs. There also need to be pull factors to the city. Redfern's explanation in part accounts for this by the fact that the older the housing the greater the potential change and investment that can be made with the surplus capital released by the affordability of domestic technologies. Other pull factors to the city are provided by Ley's (1996) new generation's pro-urbanism. Ley is careful to avoid basing the back-to-the-city movement as simply a form of what he calls 'residential credentialism' of social status arguments by suggesting that this was politics as well as status and part of a much more inclusive pro-urban movement for a liveable city. But Ley's argument does acknowledge the significance of distinct consumption habits around unique objects, bought from second-hand shops and bespoke production, as elements in a consumption realm that stands apart from the mass-produced housing and mass-produced consumer goods of the suburbs. More recently Ley (2003) suggests how the importance of the aesthetic valuation of objects by artists is a catalyst for new distinct circuits of consumption represented by gentrification.

The value of central urban land, the push factors from the suburbs and pull factors to the city combine to provide an environment in which a change in residential habits of the middle class is conceivable. The mechanism of change itself is, I argue, an emerging network of 'like minded' professionals, mostly in the public sector, with low amounts of material capital (some of them 'marginal gentrifiers'; Rose 1984) but high amounts of cultural capital whose lifestyle outlooks and social networks sustain a stable set of expectations about the desirability of buying run down housing in working class neighbourhoods (Bridge 2001a,b). What is clear from much of the research on gentrification is just how self-conscious initial stages of the process were. Early gentrifiers were staking out an alternative lifestyle from the suburban norm (in the way Ley suggests) but equally they relied on a network of others with similar assumptions. Expectations converged on a new norm of pro-urbanism that relied on social networks of others in similar situations across the city (in a sort of sparse density to adapt Wellman: see Butler 1997; Bridge 1994). These social networks become key structures of recruitment of other gentrifiers. As the process gets established other forms of economic rationality get entrained. This includes investor-developer activity, at first small scale (local builders buying and turning over properties for gentrification), through to large-scale commercial development in new-build designer neighbourhood as things really hot up. This is part of a wider economic and symbolic reordering of the central city that includes in its most advanced form a 'critical infrastructure' of discussion (including style magazines, restaurant critics and the like) that underpins the gentrification landscape (Zukin 1982, 1995).

The landscape of consumption with bars, cafés and restaurants alongside the offices of quaternary services in the post-industrial city gives gentrification the air of a consolidated social and spatial form. Whether it is sufficiently stabilized to be self-recruiting over the generations, a settled set of dispositions from which future class aesthetics will be drawn, is for me an open question. Such class stability might be confined to global cities that have a sufficiently strong high value quaternary sector labour market. The commodification of the gentrification aesthetic does seem to have given it the appearance of a consolidated class process, witness Podmore's (1998) arguments about the international portability of the loft living idea, suggesting an international gentrifier class of cosmopolites that are receptive to the idea. After all cosmopolitanism was the defining characteristic for Gouldner when he defined the 'new class' over 25 years ago (Gouldner 1979).

COMMUNICATIVE RATIONALITY AND THE GENTRIFICATION AESTHETIC: THE GENTRIFICATION PREMIUM IN INNER SYDNEY

The nature of the gentrification aesthetic in relation to economic and cultural capital was the subject of research conducted in Sydney's inner west (Bridge 1997a,b). The research explored the interrelations between taste and price in the gentrified housing market. It was an exploration of the tensions between economic rationality (narrowly understood) and the communicative rationality of aesthetic display.

There are processes that suggest continued tension in the transactional rationality of gentrification. The first is that the aesthetic/economic processes do not stand still. Gentrification has always been symptomatic of a new middle class that is so aesthetically self-reflexive. Gentrification practices have been constantly scrutinized and their very nature makes them public and visible (eating out and other aesthetic displays of cultured consumption in housing and leisure). These tensions can be seen in the relationship of economic to cultural capital. In the early stages of gentrification economic capital is small and distinction relies on significant deployments of cultural capital. At this stage aesthetic distinction can be signalled by a lick of paint (in the appropriate pastel shade of course). As the gentrification builds it involves greater amounts of economic capital to do the work that cultural capital alone was doing previously. There are now certain material entry-level requirements. The aesthetic borders of gentrification are sharply defined. Thus it is possible to over-gentrify, to make the home so authentically Victorian with antique furniture, flock wallpaper and crowded interiors that the price reached at market is severely discounted.

Economic rationality is also tied to certain dominant socio-cultural tastes. There is a connection to be made here between gentrification and an Anglo Saxon taste. This is clear from the Sydney research, where any ethnic or over-authentic renovation suffered at sale (see Bridge 2001b). In fact it was seen as unrenovated. As one Sydney estate agent reports:

> We sold a house in 'Cranston' Avenue once which was absolutely a Mediterranean home. When I say Mediterranean, it was owned by an Italian chap who spent an absolute fortune, maybe $300,000 on this beautiful home, and he'd ruined it. He'd taken the timber floors out, he'd

fully tiled the property completely throughout, he'd taken all the timber windows and put aluminium windows in, he'd put fountains and made it just out of character with Glebe – . . . because of the type of people who are buying – they said it's unrenovated – we've got to restore it – we've got to get it back to the Victorian style.

(Estate agent quoted in Bridge 1997b: 97)

In this sense a certain economic logic linked status to certain socio-cultural tastes that had the effect of eliminating difference. This can be seen in an expanded sense in the association of gentrification with whiteness. In the inner Sydney district of Redfern it has resulted in the erasure of aboriginal history in the gentrification of that district (W. Shaw 2004).

Interiors had to show the correct balance of history and modernity. Too much history is eccentric, too much modernization brings the aesthetic uncomfortably close to working class renovation activity (what Clay 1979 has called incumbent upgrading). This suggests the social freight that the gentrification aesthetic is made to bear. As Jager expresses it:

Victoriana is a fetish, in Marx's sense, in that objects of culture are made to bear the burden of a more onerous social significance, and yet retain a distinct material function. This is clearest with internal renovations, where actually the authenticity of the twentieth century working-class home was as undesirable as that of the nineteenth century Victorian home was unrealisable. For economic investment in Victoriana depended upon thoroughly modern renovations, especially in the kitchen, and the pro-vision of modern appliances. The Victorian aesthetic had its limits; it legitimates but cannot be allowed to compromise economic investment.

(Jager 1986: 84–5)

Within these hard boundaries there is a great deal of dynamism of the aesthetic. There is some evidence of a good deal of self-conscious jockeying for social position through individually distinctive embellishments on the base aesthetic. This is what I call the gentrification premium. In Sydney this meant that unrenovated properties (often deceased estates) were the most sought after. They usually had all the original features and presented a blank canvas for gentrifier conversion. The prices for unrenovated properties exceeded those for properties that had been converted structurally in the 1970s to accommodate modern amenities but that still had all the 1970s fittings and fixtures (which could have been removed for very little cost). However the unrenovated properties were in danger of costing more in terms of economic capital.

One Sydney estate agent sums up the situation:

I know one in 'Rudolph St' which was by all means derelict . . . a deceased estate, and we were looking a selling that at $(Aus) 260,000 and we achieved 350 at auction. Now at that time if that same property had been renovated we would probably have sold it for 35, 360 . . . it made the same price . . . yet to get there the person who bought it said he was going to spend 150,000 on it having spent 350 on it . . . it wasn't a 500,000 property and it wouldn't be a 500,000 property now.

(Quoted in Bridge 1997b: 98)

This suggests that the aesthetic evaluation can indeed threaten to compromise the economic investment. This gentrification premium is expressed in ever more elaborate interior redesign (some under the guidance of architects) to turn the house into an aesthetic experience of social display. In the Victorian house status was seclusion and privacy, with everyday functions, such as cooking, highly gendered and relegated to the back spaces of the dwelling. In the gentrified house status is communicated through opening up the inside of the home to the outside and through the appropriation of the outside world into the home. The highest status inclusion in Sydney is a harbour view. The structure of the Victorian terrace may undergo considerable structural renovation to capture this asset. As a Sydney agent described one house:

This is a two storey property and the outlook is from upstairs, living is upstairs, kitchen, lounge, dining and balcony is all upstairs and it's got a fantastic water view but it's just a city skyline view . . . bedrooms downstairs and your living upstairs 'cos it captures the view – that's what you impress your visitors with, the view. You don't say, yeah we've got a view, come into my bedroom have a look at that, they go into the lounge.

(Quoted in Bridge 1997b: 96)

The gentrification aesthetic as a space of identity extends beyond the home in public displays of taste and consumption in the coffee bars and restaurants of downtown chic (Bridge and Dowling 2001). It also extends into realms where the more narrowly economic and instrumental have been more prominent. In the case of the Sydney research this included properties bought by gentrifiers purely for investment purposes. The rental property had to reflect the gentrification aesthetic in a form of over-investment, even discounting for the higher status tenants it might attract and the higher rents that could be charged.

Again a Sydney agent explains in the case of houses bought with 1970s conversions:

From a rental point of view it makes no difference. It's clean, tidy, functional. Bring rents on bedrooms, not cosmetics. Certainly the conditions of the house will reflect the quality of your tenant. But if I've got a three-bedroom, two-storey house with a 70s kitchen and bathroom in it that's clean and tidy it will rent for probably ten dollars a week less than the three bedroom that has the immaculate, state-of-the art kitchen and bathroom, but the cost between the two might be 50 to 80,000 dollars difference . . . They can't get through this mentality that it's purely an investment property and there to give them a return.

(Quoted in Bridge 1997b: 98)

The status consciousness of gentrifiers even extended to their investment properties. The aesthetics of the interiors of their rental properties had to reflect their own good taste to the point where there was considerable over-capitalization on the property given its prospective rental income stream. In Veblenian terms this is a form of conspicuous consumption that even goes beyond the immediate consumption environment of the gentrifiers themselves, another manifestation of the gentrification premium.

Bourdieu's concept of habitus is 'the capacity to produce classifiable practices and works, and the capacity to differentiate and appreciate those practices and products (taste) [by which] the represented social world, i.e. the spaces of lifestyles, is constituted' (Bourdieu 1984: 170). Those with social power have a monopoly over ways of seeing and classifying objects according to their criteria of good taste. The ability to create new systems of discernment is class power. Gentrification can be seen as one such reclassification in which inner urban living became once again invested with ideas of status style and cosmopolitanism. The innovation in taste is an act of symbolic violence over others, in this case the working class residents of the inner city. This is the aesthetic border that is the equivalent of gentrification-induced displacement in those inner urban neighbourhoods themselves.

THE AESTHETIC URBAN ECONOMY

From the perspective of this chapter gentrification can be seen as an emerging and then expanding socio-cultural institution in which the instrumental rationality of economic returns are in conflict or confluence with communicative rationality based on aesthetic discernment and status display. The socio-cultural institution of gentrification is itself ever-changing. In the current period there is an emerging upper core of competition and commodification amongst elite gentrifiers in global cities. At the other extreme there is initial evidence for a more provincial form of gentrification in which consumption habits lack the distinctiveness of the earlier periods of the process and an overlapping of working class and middle class taste (Bridge 2003). In the latter case gentrification is a socio-economic movement and aesthetic institution that is worn very lightly by the residents passing through partially gentrified neighbourhoods as part of a more traditional suburban lifestyle.

What I am suggesting is that the commodities of gentrification are invested with a variable amount of weight in social communication. This raises interesting questions about the changing social 'life' of commodities. In an earlier period of urban change analysed by Walter Benjamin commodific-ation and commodity fetishism was the way in which 'commodities . . . store the fantasy energy social transformation in reified form' (Buck Morss 1989: 29). Benjamin evokes the idea of the commodity as a poetic object. In capitalism objects are not simply inert instrumentalities but rather poetic in that they connote, or are invested with, a whole range of social meanings. In this way aesthetic practices are at the very heart of the most basic element of the economy and the commodity.

Cities are very much part of this aesthetic repertoire in their ability to stage the commodity or provide new aesthetic resources to invest it with social freight. The gentrification object – the interior of the house – is com-modified not just in terms of price at market but degree of discernment required to get the right mix with individual distinction. The subtleties, the shades, are central economic 'facts' in their relation to a whole set of social processes and symbolic re-scriptings of the city that relate to economic and class power.

The importance of aesthetic evaluations instilled in practical rationality in everyday life is stressed by Bourdieu. He also suggests that they are part of wider symbolic systems in which symbolic power is exercised as a form of social division. Wealth is not just the possession of commodities but the social power that is inherent in the ability to define what objects connote aesthetically. The poetry of the commodity, as Benjamin puts it, is its openness to the ascription of symbolic meanings, meanings that more and more separate ideologically use value from exchange value. The aesthetiz-ation of the commodity is part of a process in which forms of representation are aestheticized, produced by the culture industries. Aesthetisized represent-ations of gentrification are clear, Neil Smith argues, in the scripting of the city via the discourse of the gentrification frontier (Smith 1996). This is the use of the city as a text in which certain representations prevail as a form of social power. There is more to Benjamin's argument however. The aesthetiz-ation of the commodity and the scripting of the city are part of a wider urban fantasy from which no one escapes. This is epitomized by the arcades of Paris 'the original temple of commodity capitalism' (Benjamin 1999v: 86).

The melding of desire, especially male desire, with the commodity, is at the heart of Benjamin's observations on the urban economy. The city is the shop window of capitalism. In the arcades there was a heady mix of com-modities and spectacles that transformed those walkways into another world, another city. The cornucopia of commodities fills up the senses at the same time that the sense of others is being flattened, becoming more about surface appearances. There is an atmosphere of surface reflection and mystification.

The point of this is that within capitalist relations all social ranks are caught in the desire economy. The 'anxiously modelled interiors' of the Victoriana aesthetic described by Jager through to the more self-confident 'stageings' of the home I have described as the gentrification premium, are all symptomatic of a city and an economy in which these symbolic arrangements are forever changing to sustain ever-renewed consumption.

A NEW SOCIO-ECONOMICS OF THE CITY?

An understanding from Benjamin of the commodity as a poetic object suggests how the commodity in capitalism is open to a whole range of social meanings and interpretations. It conceives of the city of spectacle in which the staging of the commodity bombards the senses and seduces the consumer as commodities take on expanded meanings through fetishization. This accords with ideas of the postmodern economy consisting of a proliferation of signs and simulacra that collapse the distinction between the original and its copies and makes consumption out of a series of images (Baudrillard 1981).

Using the example of gentrification we can imagine how the contemporary landscapes of gentrification, the recasting of the inner city as a middle class consumption landscape symbolizes the city as commodity. It suggests how the symbolic reordering of urban space can result in a landscape of continuous consumption where the fantasy energy of the commodity becomes the urban landscape itself and where the central city becomes a setting for continuous consumption of images, another Disneyland or Las Vegas strip (Zukin 1995). Similar enclaves of gentrification are increasingly found in the major cities across the world. This indeed is gentrification generalized (Smith 2002).

And yet. The lessons of institutionalism are that economies are continuously evolving and unpredictable and that the linear path of aestheticization and commodification is not inevitable. The economies of cities are indeed socially and culturally embedded (Granovetter 1985) and, to take the emblematic example I have used throughout this chapter, this can lead to different manifestations and fates of gentrification. The particular constellation of factors in place means that place matters and is a vital component of the ongoing socio-cultural development of economic institutions. My research in London and Sydney has identified a gentrification premium. Here the central city itself has become a positional good, one that communicates symbolically as well as materially the status of its residents. In Sydney the gentrification aesthetic has intensified to include confidently modelled interiors of houses, often architect-designed with a premium on view of downtown or the harbour bridge. The aesthetic premium extends not just to the lived environment of gentrifiers but to their investment properties where renovations are to specifications and aesthetic standards that far exceed instrumentally economic requirements given the type of rental returns. The aesthetic requirements extend to the performance of bodies in public in forms of tasteful self-presentation and ease of movement and interaction in the bars and restaurants of gentrified neighbourhoods. The performance of bodies and these spaces become themselves the subject of reflexive discussion via the critical infrastructure of media and culinary reviews (Zukin 1995).

But the gentrification premium is not the inevitable fate of all cities that experience gentrification. There is gentrification that is more diffuse, both in terms of its impact on the neighbourhoods in which it is found and in terms of the self-reflexive agency of the gentrifiers involved. My recent research in the provincial city of Bristol, UK, suggests diffuse gentrification (Bridge 2003). Bristol has a postindustrial employment structure, with a strong represent-ation of financial services industries (many decentralized from London), creative and media outlets. It has waterfront designer neighbourhoods as well as the sweat equity Victorian suburbs. Despite having many of the ingredients of premium gentrification the research suggests a gentrification process that reflects a provincial city and the particularities of the housing market of Bristol. The neighbourhood studied is characterized by continued social mix and the aesthetic self-consciousness of the gentrifiers is at a much lower level than the gentrification literature generally implies. The con-spicuous aspects of the gentrification did not figure strongly and there was evidence to suggest an overlap in aesthetic terms between working class and middle class residents of the younger adult cohorts. This lack of distinction, and a more generalized consumption aesthetic, is noted by Wynne and O'Connor (1998) in their study of gentrification of inner Manchester. Here the contemporary music scene provided an aesthetic realm that blurred distinctions between working class and middle class taste. Contemporary music was very much part of the urban scene, part of the way the city was appraised and consumed.

As well as suggesting the blurring of the gentrification habitus across traditional boundaries of class, the Bristol research gives some indication of its dissipation further down the lifecycle and housing trajectories of putative gentrifiers. Concentrating on the lifecourse of gentrifiers it became clear that the aesthetic practices of gentrification were often trumped by more tradi-tional middle class concerns such as access to good schools. Gentrifiers interviewed prior to and after moving out of the neighbourhood (or within the neighbourhood) showed a wide range of destinations in terms of both location and the aesthetics of the housing (from new build suburban to Georgian villa). The part-gentrified neighbourhood studied is just one route in a range of middle class housing moves across the city and region and not vital to the constitution of a new middle class identity.

The particularities of place are part of the socio-cultural embedding of markets that make for a great deal of variability and distinct geographies of gentrification. These geographies are socio-cultural complexes that show variability in the 'value' of gentrification. This contrasts with the conception of value as an invariant, both from neoclassical theory (involving immutable 'utility') and the Marxist idea of the labour theory of value. Rather than bid rent or the rent gap triggering the process of gentrification with stage models from lick of paint sweat equity to corporate development of designer neighbourhoods, there are a range of socio/cultural/economic institutions of gentrification out of which different valuations of the process emerge. There is no essential value which builds up via aestheticization and commodification, but rather a range of valuations that have different potentials. Like all economic institutions gentrification stabilizes the definition of a commodity – in this case the relative social and monetary desirability of inner urban land and buildings and its measurement or relative intensity, usually measured in monetary terms. The way that gentrification has built up over the last 40 years, from the conversion of the little artisans' cottages in London, to becoming a vital component in global neoliberal urbanism (Smith 2002; Wyly and Hammel 2004) is surely testament to the forces of universal commodification and an ever-expanding value sphere. But gentrification can also be understood as a range of evolving socio-cultural complexes in which the social constitution of value varies. There is no inevitability to gentrification generalized. At one extreme there is the designer neighbourhood of the global city that combines a work and leisure landscape of stylized deportment and consumption. This is the exclusive socio-cultural complex that is a form of cosmopolitanism. The pressures of gentrification here are intensifying in a spiralling set of competitive practices at the leading edge of taste. At the other extreme there are socio-cultural complexes of gentrification that are diffuse and dissipated. Here the aesthetic practices and body dispositions do not connote distinction.

The different socio-cultural complexes of gentrification also impact on the degree to which social division and displacement is evident, as well as the ability of different institutions to make a difference to the way in which it works out. As K. Shaw (2004) has suggested for St Kilda's in Melbourne, local circumstances do matter – in this case in the ability of the local state to control gentrification. The ability to resist gentrification, or make its effects more equitable, depends on the strength and coherence of the transactions in the particular socio-cultural complex of gentrification. In the next chapter I consider how planning can influence these processes as well as how the idea of the rationality of planning itself is changing.

7 In city hall

Planning theory has been vitally implicated in ideas of rationality. Indeed forms of rationality stand at the heart of two of its main approaches – the rational comprehensive planning model and the communicative rationality model. Radical planning (including Marxist political economy) sees the rational comprehensive model as hopelessly positivistic and the communicative model as hopelessly naive about structures of power (and both meliorative of capitalism). Rationality is still required by the radical model, however, both to demonstrate the laws of motion of capital and to oppose it with strategically rational social mobilizations. Postmodern planning on the other hand abandons rationality altogether as hopelessly modernist and conferring a logic of identity that denies difference. In this chapter I first discuss the changing relationship of planning to rationality. What I then suggest is the significance of pragmatist planning theory – post-postmodernism. Indeed I argue that planning theory has not been pragmatist enough. Although it has taken us to a consideration of non-discursive as well as discursive communication (Forester 1989, 2000), that communication is still seen in representative and consensual terms. If communication is viewed agonistically (within communities, and lifeworlds and within voices as well as between them) and as a form of *argumentation,* we have a much more perspectival view that sustains difference. Understanding these tensions in discourse starts to tackle some of the criticisms that pragmatism is too naive about power and structure. It also addresses critiques, such as Sandercock's (1998), that suggest that it is not able to account for change and transformation. Although respectful of difference, unlike postmodern planning, it retains a vital role for communication across (as well as within) difference and the significance of rationality in that endeavour. Attempts to harmonize agonistic discourse is one facet of pragmatic rationality, and one that leaves us with a very different view of the relationship between planning and urban space. It also suggests a much stronger role for planning than either the postmodern or Marxist prospectuses allow.

THE RATIONAL COMPREHENSIVE PLAN

> Planning was a rational procedure, which was above the economic conflicts of the American city; it paid more attention to the meshing of components, the adaptation of parts, and the procuring of efficiency; it directed attention to the simplistic relations and elements that could be regulated, such as the flow of traffic and the disorderly and conflicting arrangements of land uses. Allowing the exploitative processes of capitalism to occur behind their backs, city planners thus introduced an ideal form into their theories, an abstract rationalization that made possible the removal of some of the barriers that thwarted economic growth yet protected capitalism. As upholder of the ideal solution, it was the aim of planning to be purchased by a growing body of public administrators and voluntarily adopted by private property owners, to be sold as partial reform realized through efficiency techniques that transformed obsolete land uses into more productive values, provided new infrastructural requirements for capital investments and accumulation, and maintained a harmony of social consensus and legitimation. This is what I have referred to as the planning mentality.
>
> (Boyer 1983: 69)

Christine Boyer's analysis focuses on American city planning but her arguments extend to the planning mentality in most western cities in the modern era (Boyer 1983; but also Hall 1988, 1998). What she describes we might refer to as high modernist planning, based on positivistic knowledge. It means that reality can be broken down into isolated component parts. Identifying the relationships between the parts explained how the world worked. Technical interventions in these relationships would alter the city system and make it more efficient (usually, as Boyer suggests, for the smooth running of the capitalist economy – either deliberately or inadvertently). These assumptions meshed with a diverse set of influences from the earlier utopian modernist phase of planning, seen especially in the work of Le Corbusier, Frank Lloyd Wright and Ebenezer Howard. Utopian and high modernism in planning had a number of things in common – assumptions of control, completeness, efficiency and a particular conception of urban space. These all characterize a dominant form of instrumental rationality in the planning imaginary.

The sense of control was the idea of planning as an intervention that helped discipline or contain some of the more unruly forces in the city. Early philanthropic interventions (in health and education) were aimed at correcting and containing some of the worst effects of the industrial city on its poorest inhabitants. This persists in the idea of development control. As the planning profession got established in the late nineteenth and early twentieth centuries control came more and more to mean the technical ability to predict and intervene. That was tied to control as professional autonomy, based on specialist knowledge and the application of particular skills. Professional knowledge involving analysis and prediction, and practical application of that knowledge as a form of disciplinary control, helped establish the separateness of the planner. This was elite knowledge. The planner was above the fray – he (and it usually was he) was able to observe the city as a whole and intervene dispassionately to correct some of its unruly energies.

Professional control meant a completeness of view. Inherent in the comprehensive master plan this was the assumed ability to see the city as a whole (either as a body, or a machine) and to be able to predict what interventions in one part would do to the other parts. It is the ability to 'see' the city, to treat it as what Lefebvre (1991) would call a visible-readable realm. Visualizing the city also implies a separation, an ability to envision whilst not being embroiled. Visuality is cool, calculating and judgemental. Anthony Vidler has shown the importance of the Le Corbusier's flights over Paris and the use of aerial photography for his master plans of the city and its buildings as machines for living in (Vidler 2000). Planners and architects not only had specialist knowledge but a privileged view of the city, separate, above, godlike.

The completeness of view of the city as a body or machine also implied an idea of its smooth running. With a utilitarian heritage and the influence of the maximizing model from neoclassical economics, smooth running meant instrumental efficiency, maximum output for minimum cost. The dominance of means–ends efficiency and instrumental rationality in planning (and public administration more widely) had a number of effects. One was to regard ends, or benefits, in terms of economic output. This narrow conception of efficiency focused on technical means for economic ends, but at the expense of any wider value considerations. This is the separation between value and purposive instrumental rationality of which Max Weber (1968) forewarned.

Instrumental rationality in the planning imagination is what Lefebvre describes evocatively as 'a narrow, desiccated rationality' (Lefebvre 1991: 200). It also has a particular relation to urban space. It is applied to 'representations of space', which in the planning imaginary are tied to Cartesian space, a two-dimensional geometry with points and intersecting lines – abstract, visible, readable, abiding by a logic of mathematics. This is a space of mental abstraction rather than material experience. It supports the Cartesian dualism of mind and body. The mind is a separate realm that represents materiality (and the body) in certain ways. Abstract visualization dominates the lived materiality of the city, according to Lefebvre. This is supported in the planning imagination by a conception of space emptied out of its material content and operating as a two-dimensional space of the balance sheet of instrumental profit and loss. It separates from, and therefore is able to support, the materialities of capitalist production. It is a conception of space that abstracts away from, and then subsumes, difference.

All in all the modernist planning imaginary is based on a rationality that means the integration of form and function in a city considered as a whole integration of activities. It is a rational ordering that relies on the specialization and separation of spaces and their integration through efficient circulation (of traffic and communication).

THE RATIONAL COMPREHENSIVE PLAN: THE BUILDING OF BRASILIA

In his magisterial analysis of the building of Brasilia, the new capital of Brazil built in the 1950s and 1960s (and which drew heavily on modernist architectural and planning guidelines) Holston (1989) argues that the modernist system is based on the integration of elements of equivalence, rather than difference.

Holston gives the detailed prescriptions within the zoning model that demonstrate the comprehensive, functional aspect of the organization of residential neighbourhoods, as if they were viewed from above.

> The function of residence is typologically opposed to that of work by the arms of the axial cross of speedways: the Monumental Axis of work sectors divides the Residential Axis into its South Wing and its North Wing. In mirror image, these organize the quotidian and domestic domain of residential functions. Each wing is divided into nine bands (fixas), numbered 100 to 900. The band's number indicates its position to the east (even) or west (odd) of the axial speedway. It also indicates its function. Band 500 is a commercial strip called W-3 that also contains community facilities and services (such as libraries and clinics) at the entrances to the 300 band of superquadras [residential superblocks] between W-3 and W-2 . . . each band in the 100 to 400 series contains 15 superquadras yielding a total of 60 in each wing and 120 in the entire city.
>
> (Holston 1989: 166).

And so on. Holston shows how this abstract and totalizing conception of the city in the planning imagination influences the way that Brasilienses mentally represent their city. Each sector is typologized in terms of form and function and each sector fits within the overall cross shape of the city inscribed by the fast roads (speedways). As Holston describes it:

> As we might expect, its total order inverts the problems of orientation associated with other cities. It increases the legibility of the whole but decreases that of its parts, producing a peculiar set of navigational dilemmas. Most cities are not associated with total shape. They present a non-figural conglomeration of sprawling districts within which, however, individual neighbourhoods are identified by distinctive landmarks of

one sort or another . . . in contrast Brisilienses understand Brasilia as a single, legible image – commonly read as a cross, an airplane, or a bird – composed of neighbourhood units that with very few exceptions they find uniform, undistinguishable, and landmarkless . . . When one asks them for directions, for example, they will inevitably reckon by the whole first, describing the cross in some fashion and then locating the desired point within it. Or, they will simply give the address, which again depends on a knowledge of the whole. Both modes of reckoning are entirely abstract.

(Holston 1989: 148–9)

Yet Holston goes on to describe ways that the inhabitants of Brasilia have appropriated this space of modernist rationality in different ways – how they have inscribed difference on the geometry of abstract space. One example is the response to the death of the street. Circulatory routes have eliminated street corners, a traditional site of male sociability in Brazil. The reassertion of street life is seen most particularly in the commercial sectors. Here the store-owners have reversed the design of the shops, abandoning the front entrances through garden space for the rear entrance which abuts the curb and recreates the sense of a busy market street typical of other Brazilian cities. Many wealthier residents rejected the egalitarian designs of the super-blocks of apartments (the superquadras) and created their own exclusive neighbourhoods of detached houses outside the Plano Piloto. These houses contrast with the modernist uniformity of the international style superblocks with a bricolage of historical styles.

Holston also describes the way that modernist comprehensive, rational planning has inadvertently destroyed the public realm in Brasilia by designs that were meant to free up the city's spaces to encourage access for all. The large green spaces between the superhighways and around the superquadras are often only accessible by car, are large, open and inhospitable, and do not combine the variation of use and design found on the traditional Brazilian street. As a consequence social life has been interiorized, specialized and segregated to a much greater degree.

Holston's study of the building of Brasilia demonstrates several elements of the rational comprehensive plan, including its abstractness, its use of Cartesian geometry and desire for functional integration of the city as system. It also shows how these abstractions are often at odds with the lifeworld orientations of ordinary residents and also how space can, in some small degree, be reappropriated by those without power.

The modernist planning imaginary still dominates planning practice but there have been a number of recent developments in theory and practice that start to resist this dominance. I concentrate here on four major rationality paradigms that underlie these models. The first (already considered) is the comprehensive rational model that works on an instrumentalist basis, using technical criteria, in an attempt to compensate for capitalism. The second, radical planning, maintains an instrumentalist component but works on a radically different space, a non-capitalist space. The third, deliberative planning, is critical of capitalism but accepts the necessity to work within it. It does however have an alternative idea of rationality that we might call deliberative, communicative or participatory. This latter model shades into a fourth idea of rationality in planning, which is fully pragmatist.

THE 'DELIBERATIVE TURN' IN PLANNING THEORY AND PRACTICE

John Forester's (1989, 2000) detailed analysis of the practice of planning reveals that in many circumstances the rational comprehensive model is found wanting, either as a planning heuristic, or indeed as a description of how planners act. Rather than being the distant elite experts who rarely venture out of city hall, Forester sees planning practice as deeply embroiled in uncertainties and emotional conflicts and much more involved in the lives of those the planning initiative would affect. Planners never have what we could consider to be perfect information, knowing all the options and being able to make clear-cut decisions about clearly defined problems, as the rational comprehensive model suggests. Forester first explores these limitations in ways that economists have done – through the idea of bounded rationality (Simon 1957). Simon's model replaces the idea of optimizing with satisficing in decision-making. In uncertain conditions and lacking perfect information decision-makers may take the satisfactory option (in terms of aspirations), rather than seeking to fully maximize their utility. Forester's adaptation of this model sees the constraint in terms of limits to the computational abilities of the planner: the fact that the planning situation is socially differentiated (involving different actors with their own interpretations of the situation). The latter constraint makes satisficing difficult – as planners have to exploit social networks in a form of information gathering to help inform the decision. A further constraint is that of pluralist conflict: the fact that different actors have different interests means that the political perspective that informs other's interests and opinions has to be taken into account. Structural distortions are the fourth constraints on the decision-making situation and these comprise structural inequalities in power between the actors in the situation.

The bounded rationality model is in some senses exploring the informational constraints on the rational comprehensive model. But what these constraints reveal is the fact that the status of information is conditioned strongly by the context out of which it emerges (see Innes 1998). It also suggests that these competing perspectives in reality are not easily resolved and that actually much of the activity of the planner is absorbed by the procedures of familiarization with competing worldviews and power resources in the process of coming to a judgement about how to proceed. This contrasts with the assumptions of the rational comprehensive model that takes these factors as assumed informational constraints that are post hoc and unexplored. The move to a greater weight attached to the procedures for understanding the situation into which the planner enters to intervene takes Forester in the direction of deliberative, or communicative rationality.

Forester and others have drawn on the philosophical work of Jurgen Habermas to make sense of this. Habermas's (1984, 1987) idea of communicative rationality suggests a very different operation of planning rationality in the city. It also projects a different idea of the objects of planning. As we have seen in Chapter 5, Habermas's defence of rationality involves understanding it as a social and linguistic competence, rather than the cogitation of individual minds. Its criterion of success is not instrumental efficiency but communicative understanding. Habermas is arguing that the normal, everyday structures of language contain mechanisms for conveying meaning that rely implicitly on validity criteria. These validity criteria relate to a person's objective knowledge of the world, their social norms and their subjective views. Each of these realms of knowledge has its own validity criteria – truthfulness, legitimacy and sincerity – respectively. Participants in conversation speak from the situated context of their objective, social and subjective knowledges but in the knowledge that their validity claims might be redeemed or challenged by other partners in conversation at any point. This brings a deliberative element into the exchange of meaning because participants must examine the underlying validity claims of their statements in order to defend them. This defence is made on the basis of the good reasons or grounds for the statements. Good reasons or grounds are rational criteria for a procedural defence of situated validity claims – that is communicative rationality.

These detailed philosophical points have a wider resonance in terms of our understanding of urban planning and the city more widely. In terms of the defining criteria of high modernist planning it works as a critique of control, completeness, efficiency and the conception of urban space. Habermas uses the insights of speech act theory to make the separation between communication aimed at exchanging meaning (locution and illocution) and communication that aims to get others to do something (perlocution). In capitalism perlocution is pervasive and tied to instrumental rationality, using others as means to ones own ends. It is the rationality of the system. Communicative rationality is based on the communication of meaning for mutual understanding, a competence fostered in everyday and intimate settings – in the lifeworld.

Planning based on communicative rationality seeks to understand the lifeworld context of individuals and groups. It sees planning as a dialogue to try to understand lifeworld meanings such that any planning intervention is more sensitive to context, and to the way others see their world. Planning is like a conversation rather than an intervention. It is a conversation in which the planners must also defend their validity claims – in terms of their objective, expert knowledge – but also their own social norms (in fact social norms and the norms of their profession) as well as their subjective views of the situation. This means a lessening of expert control in that expert and lay knowledge (of both planner and community) are brought closer together. Furthermore the modernist idea of completeness gives way to an appreciation of the messiness of most socio-physical contexts in which planning decisions must be made. Efficiency in context can no longer claim the abstract qualities of narrow instrumentalism but must be cognizant of situated knowledges and value systems. If the rational comprehensive model of planning assumed a universality of expert knowledge applied to different situations, the deliberative turn denies the universality of content of planning but defends a universalism of deliberative procedure to come to agreement over planning issues. This is a consensus-based, collaborative idea of planning (Healey 1997).

Listening to validity claims is crucial to this process for Forester. Listening means being attentive not just to what people say but to the potentially deeper messages behind what they are saying. It involves careful attention to the exchange of speech acts that are based on validity claims of comprehensibility, sincerity, legitimacy and truth. It is 'critical' in that it is aware that communication is not necessarily mutually enhancing but can be distorted in systematic ways as a result of power and sectional interest. In fact Forester lays out the ways in which communication can be distorted in terms of comprehensibility, sincerity legitimacy and truth at the level of face-to-face interaction, organizations and the political economic structure.

THE OTHER SIDE OF RATIONALITY: RATIONALITY AS RATIONALISATION

In Forester's model power can distort communication to undermine the rationality of action (in this case planning interventions). In a book impressive for its empirical detail and theoretical sweep, Bent Flyvbjerg (1998) offers a detailed case study of such communicative distortions. In his Foucauldian analysis of the planning of a central city bus station in the Danish town of Aalborg, Flyvbjerg shows how the procedures of rational deliberation mask the operation of power and sectional interest. This meant that the Aalborg project failed to meet wider democratic expectations. Power trumps rationality and those with power, the commercial interests and the city council transform rationality into rationalization. Rationality only exists in relatively stable situations, and because most planning situations are confrontational, rationality is relatively weak.

There are a range of competing interpretations as to whether Flyvbjerg's case study of the Aalborg project shows the limitations of the ideals of deliberative planning or merely the considerable obstacles that such communicative distortions present for communicative rationality. Flyvbjerg's closely observed case study reveals the operation of rationalization, a systematic distortion of communication (in Habermasian terms) in which rationality operates not to uncover the evidence in an even-handed way but to serve sectional interests of the powerful.

In the Aalborg project one example of this is the confrontation between the office of the city architect and the bus company over the size and location of the new bus terminal. In this case the architect and the technical department adapt to the bus company's demand for a larger, more centrally placed terminal. In interviews with the task force members Flyvbjerg reveals just how this happened.

> Interviewer: In the Task Force you all agreed to propose Nytorv [central city location for the bus station] to the Executive Committee. At the same time you decided that a memorandum had to be written which should describe the advantages and disadvantages. In terms of a normal logical sequence, I would think that it should have been the reverse.
>
> Task Force Member: Yes, that's entirely clear. The problem in planning is time pressure. There is, in fact, a set schedule for when a bus terminal at Nytorv-Osteraagade must be established. And whether starting to look at the other [location possibilities] is a plot, or what it really is, I think can be questioned. The memorandum we wrote lacks a planning and traffic related evaluation of the different location options, for we had no chance to carry out such things.
>
> Interviewer: But isn't this just good old fashioned manipulation? To make a decision on something and then say 'Let's make a memorandum which shows that it's a right decision.'
>
> Task force Member: I wouldn't say that. But you know, it might look kind of strange, right? [Laughter] And that's correct – it might, well – of course it resembles – hmmm, you know – rationalisation, right?
>
> (Flyvbjerg 1998: 24)

This example shows clearly how what gets talked about and the way it is discussed, the agenda setting, structures the dialogue in ways that restrict the possible extent of communicative deliberation from the outset.

Nevertheless there are many instances from Forester's work, and even a few from Flyvbjerg's, that suggest that power and rationalization are not complete in their relationship to rationality. The first is from Flyvbjerg himself who suggests that rationality is still a resource if only for the relatively powerless. The reasoned arguments of the representative organization for cyclists are the example in the Aalborg case. They are not able to examine the hidden rationalizations of the developers, retailers and the council, nevertheless their interventions do provide evidence of some resistance to rationalization. Forester himself also provides numerous examples of how a more deliberative approach to planning interventions has had a positive influence, even in situations of extreme confrontation that Flyvbjerg would judge to be no-brainers for any kind of rational deliberation. For Flyvbjerg rationality is a possession of the weak and all too often confronts, or is subtly distorted, by the operations of power.

In contrast to Flyvbjerg, Hoch (1996) uses Rorty's argument that Foucault reaches an analytical impasse because he (Foucault) looks to public institutions to provide the sorts of transformation that can only really be achieved by poetic self-expression. 'Planners, as public servants, would do well to leave Foucault at home (Rorty 1989, 64–6) and to carry Dewey with them' (Hoch 1996: 36). Despite the individuating nature of modern life, ideas of autonomy can still be understood in terms of communities (however diverse or overlapping) that foster the individual, understandings of a 'robust pluralism that does justice to the tangled quality of our experience' (Bernstein 1986: 18, cited in Hoch 1996: 31).

Another way of looking at this is to take a pragmatic understanding of discourse that sees conflict and diversity *within* communicative action. A wider view of communicative action also starts to question the comprehensiveness of discourses of power over the social body. Part of the reason that Flyvbjerg sees the inevitability of rationalization is that, like Habermas, he concentrates on linguistic communication (talk and texts). Non-verbal communication, emotion and affect, an altogether broader view of communication, might, I argue below, provide instances of non-rationalized potential.

TOWARDS PRAGMATIST PLANNING

Forester's work is taking the consideration of rationality via Habermasian deliberation much more into the terrain of full square pragmatism. He presents an example of the kind of rationality that can operate in the service of the disadvantaged but that is not necessarily built in an environment of consensus. He uses the case study from earlier work (Forester and Chu 1990) of the need of Allan Isbitz, a planner, to hire a homebuyer counsellor for a settlement programme in an Appalachian community with a split between the poor whites and African-Americans. The person qualified for the job to advise families before they went to the bank to try to purchase a home happened to be black and Isbitz was warned that the Appalachian poor whites would never work with an African-American person.

However the appointee brought her own practical knowledge as well as professional knowledge to their situation:

> a lot of people in fact wouldn't work with us in the beginning. But ultimately, they did. She was good, sensitive, and not overly aggressive. In fact, she was very quiet. In her own quiet way she plugged along, helped everybody over the two years, and became a very trusted member of the Appalachian community.
>
> (Forester 2000: 50–1)

Forester interprets the kind of rationality at work in this example:

. . . a kind of practical rationality that does not promise grand strategy but responds to particular need, that does not so much fix clearly on an end and choose a means to it as it responds without creating new problems, in a way, Allan tells us, 'that is not heavy handed'

(Forester 2000: 51)

He also points to the importance of emotions, both in terms of trepidation about the project as well as 'the counsellor's success [grows] not just from her competent provision of information about bank practices but from her particular emotional qualities too: her persistence, her sensitivity . . .' (Forester 2000: 53). Her own experience of poverty and her personal qualities were highly significant in moving things forward. Intellect and emotion are part of good deliberation.

To deliberate well – which requires both sympathy and detachment – one must . . . be able not only to think clearly but to feel in certain ways as well. The person who shows good judgement in deliberation will thus be marked as much by his affective dispositions as by his intellectual powers.

(Kronman 1987: 858, cited in Forester 2000: 54)

In Allan Isbitz's case it suggests some sensitivity to the situation that included a judgement of human capacities to appreciate the effort. As Forester concludes:

Allan's story suggests no weighing of commensurable costs and benefits. It suggests instead a careful work of practical rationality, a careful consideration of general facts, including racism and poverty, that would influence the actions of particular people with particular sensitivities and histories and which would, in the specific case, be influenced in turn. This rationality, of Allan's and the counsellor's, is interpretative and practically pitched, responsive to the ambiguity and incommensurability of real situations, and this rationality is conveyed powerfully by practice stories, whose messiness and moral entanglements characterize the work that practically rational actors must do all the time.

(Forester 2000: 56–7)

There are a number of issues for our consideration of rationality here. Although the kinds of judgements being described are not necessarily incommensurable with strategic rationality, they do lead us into the realm of emotional sensitivities. The sorts of assessments involved draw not just on what people say but also the performative elements of their speech and non-verbal behaviour. The elements of communication within and beyond dialogue may well be highly significant in successfully transforming the situation. What I have hitherto called 'the margins of communication' Forester sees as 'participatory rituals' and suggests that they provide participants not only with dialogue and argument, but with more of relevance than they anticipate, with more of value than they first appreciate, with possible relations with others they could not foresee, and with a literally surprising deliberative political rationality far richer than accounts of decision-making rationality or rational choice allow.

(Forester 2000: 116)

Participatory rituals are:

encounters that enable participants to develop more familiar relationships or to learn about one another before solving the problems they face – for example, the informal drink before negotiations, the meals during focused workshops; the small break-out groups complementing plenary problem solving sessions; the early story-telling phases of mediation processes, and so on.

(Forester 2000: 131–2)

As we have seen in Chapter 5 these peripheral elements of dialogue are also assumed to be significant for the more portentous exchanges by Iris Young when she stresses the significance of storytelling and greeting as important elements in the attempts to renew the public realm.

Rather than argument resulting from given interests of actors with fixed identities Forester points to the transformative possibilities of safe participatory rituals. These include changing relationships and identities, transforming issues, agendas and ends and the usefulness of structured unpredictability. In these situations participants take 'certain stances of tone and posture and attentiveness' (Forester 2000: 147; also Throgmorton 1996; Healey and Hillier 1996). Creativity and play are often important in complex and uncertain decision-making contexts (Cohen *et al.* 1988). Sometimes deliberate distraction is necessary to prevent participants focusing too narrowly or too early on their own instrumental interests.

'Decision making involves not only cognitive choice but social expression, social articulation that divisively or redemptively addresses those served and those disserved, winners an losers alike' (Forester 2000: 150). Social expression includes rituals to save face and preserve the face of others as the rituals of encounter on the street demonstrated (Chapter 3). Rationality here may be as much about the ongoing social context of antagonism or consensus, rather than the explicit content of those (dis)agreements.

The analysis of learning through deliberative, participatory rituals suggests that we learn not only with our ears but with our eyes and hearts. We learn not only from surprising information that leads us to propose new hypothetical lines of action to test, but we learn from style and passion and allusion too. We learn to reframe our predictions and strategies, but we learn to develop new relationships and even senses of ourselves as well. In participatory processes, we not only generate arguments, but we construct networks and new organisational forms as well.

(Forester 2000: 151)

New networks and organizational forms may be highly significant in the formation of political identity such that people can participate in a wider public realm. Forester (2000: 207–11) describes Mary Jo Dudley's (1996) work with domestic workers in Colombia. Accounts of the trauma of sexual harassment and abuse at work emerged as the workers interviewed each other. The experience changed their senses of themselves and each other. Their questioning went on to address wider public knowledge and attitudes. This transformation resonates with Nancy Fraser's (1992) idea of 'subaltern counter publics', as safe spaces for the realization of identity and shared interests. These spaces also have a strategic element in being a place from which to conduct wider politics.

Dudley's example shows how emotion can be at the heart of realization and mobilization.

In planning practice, fact and feeling, reason and emotion are often tightly intertwined. Whether a long-time neighbourhood resident faces unwelcome change or a developer faces financial risk, anger and fear are always close at hand. In such situations, planners must deal as much with people's passions as with their earnestly certain predictions.

(Forester 1989: 107)

Dealing with emotions as part of wider democratic engagement has long been established in neighbour mediation, and by local authorities dealing with different community interests, examples acknowledged by Hoggett and Thompson (2002) in their discussion of the burgeoning idea of a democracy of the emotions.

PRAGMATIST RATIONALITY AND PLANNING 1: LEARNING FROM SOCIAL LEARNING

The reformist, deliberative model of planning practice shades into a broader pragmatist paradigm in planning. The idea of linking pragmatist philosophy and planning practice has been established for some time (Hoch 1984a,b). This pushes planning practice even further into contextual situation. It makes the planner and the community mutually constitutive of planning practice. It combines elements of the deliberative and the radical traditions. It has a heritage, however, that pre-dates these developments. Its strongest representation came not in planning but in the social development model. Friedmann (1987) defines the social learning model as one that moves from:

... anticipatory decision-making to action and social practice. The social learning approach works with a process concept of knowledge: its central assumption is that all effective learning comes from the experience of changing reality. As a form of knowing, it is intrinsically related to human activity, focusing attention on dynamic social processes (as opposed to pure cognition), examining problems from the perspective of an actor actually engaged in practice, highlighting the mediating role of small groups, using the concept of social environment as a major category in the analysis of learning situations, and emphasizing the crucial importance of dialogue for social practice.

(Friedmann 1987: 216–17)

Healey (1997) summarizes Schon's (1983) development of the social learning model as:

1. People learn through doing: they develop theories in action. It would help the learning processes if people were more conscious of this learning in action, if they become reflexive.
2. There are two dimensions to such learning; the first, or single loop learning, involved working out how to perform a task better within given parameters; the second involves learning about the parameters and thereby changing the conditions under which tasks are performed. Schon's model of the reflective practitioner emphasises double loop learning.
3. Such double loop learning can take place in social situations, through dialogue, through which people can collectively explore and learn about issues and each other's attitudes to them. Such group dialogue can be assisted by techniques of group dialogue facilitation.
4. Problems and objectives, facts and values, emerge through such group processes; they are not waiting 'out there' to be discovered by group enquiry.
5. Group discussion processes which reach double loop learning can re-set parameters for subsequent action, and in this way engage in setting the 'frame' for action. This framing work can be equated with deliberative strategy making.

(Healey 1997: 257–8)

This represents a shift from managerial understanding of suitable knowledge to social ones, according to Healey. But the approach is still individualist with an objectivist view of the external world. It is with deliberative, or communicative planning that the emphasis is put on the intersubjective context of planning.

The social learning model is based on Dewey's ideas of the relationship between organisms and their environment that we have explored elsewhere in this book (see especially Chapter 2). Friedmann identifies two major shortcomings of this prospectus. The first is its rationalism: that human activity consists of constant problem solving and forward motion (in a sort of pale imitation of science). Habits, interests and other familiarities might be sustained for their own sakes. As I have argued this conserving energy is fully explored by Bourdieu in the habitus or weight of social practices on present action. The other issue is the validation problem. For Dewey valid knowledge disposes of a problem. Knowledge is valid if people can agree on it. Friedmann reminds us that the world is full of intractable 'wicked' problems (Rittel and Webber 1973) that are not disposed of but merely displaced by other more pressing problems. Validity in dialogue of course is at the heart of Habermas's concerns and the charge levelled here is that the form of consensus that Habermas points to is possible only in small group settings. In the wider socio-political domain this consensual rationality comes up against naked power in the form of rationalization, distortion and brute force.

RADICAL PLANNING

Confronting naked power and inequality has of course been the intent of radical planning theory and practice in its various forms. Marxist political economy has been deployed powerfully to understand the role of cities, or rather the urban process (Harvey 1978) in capitalism (Harvey 1973, 1985; Castells 1977). That role is to ensure the smooth circulation of capital, cope with any crises of over-accumulation through switching capital between circuits of accumulation. Cities are furthermore centres for the exploitation of labour and the coordination of that exploitation elsewhere. Planning initiatives that seek to counter the injustices of capitalism are to be welcomed, as are grassroots social mobilizations (Castells 1983). On the whole though, planning is seen as an arm of the state and ultimately functional for capital. Planning helps provide infrastructure and the sorts of efficient urban spaces that privileges capital and ensures its smooth running. Whereas perhaps in the past planning had a role of compensating for some of the excesses of capitalism, in current neoliberal times planning is entirely in the service of market imperatives. Planning practice has a very limited role and would necessarily have to be implicated in comprehensive social and economic transformation to have any greater efficacy in terms of social justice.

The seemingly overwhelming power of capital does not stop efforts in the planning community at trying to promote social and economic transformation. There are a range of approaches here from developments of advocacy planning (Heskin 1991; Leavitt 1994; Peattie 1994), from feminism (Leavitt 1994; Hayden 1980; Watson 1988) and struggles over civil rights (King 1981; Leavitt and Saegert 1990) and international development (Friedmann 1992) – for a review see Sandercock (1998). These movements are much more concerned with community mobilization and seeing the planner not as the representative of the community but as an agent in the community, although

Friedmann (1987) has cautioned that planners need to keep a critical distance to be effective. This has placed many radical planners in conflict with the state and adopting a much more action-based rather than theoretical and rational approach to social problems.

Although often in conflict with the rational comprehensive model, and despite being immersed in community issues, there is much of radical planning that follows the same logic of action as the instrumental rationality of the market. Radical mobilizations, with different degrees of success, are aimed at strategic mobilizations (that is strategically rational) in the sense that they seek to take on the logic of power head on, and to fight on the same ground, to fight fire with fire as it were.

In urban planning, no less than elsewhere in social theory, approaches that privileged the economic as the main locus of power, class relations as the most important social relation and versions of strategic rationality as the action framework came under increasing criticism. All these criticisms reach their height in postmodernism and postmodernist planning theory.

POSTMODERN PLANNING

In an important article Beauregard (1989) found that American planning had isolated itself from many of the debates in social and cultural theory, especially in relation to postmodernism. Planners were still caught in a modernist mindset when many of the cultural and social practices that postmodernism sought to capture were shaping the city in different ways, ways that planners should become sensitive to, both in theory and practice. More recently Sandercock (1998) calls for postmodern practice in planning. This involves a move away from an emphasis on technical expertise and instrumental rationality concerned with means to ends (though this is still required) towards a communicative rationality that focuses on goal setting. It requires a move from comprehensive planning to that based on negotiation and particularities of people and place. It also involves a shift from an overarching scientific, positivist epistemology to a range of epistemologies emerging from hermeneutics, feminism, action research and other ways of knowing. It is embroiled in community empowerment rather than simply bestowing professional expertise or engaging in limited community consultation. Planning should be seen to operate not in the name of a taken-for-granted (but power-riven) single 'public interest' but should recognize multiple publics and the politics thereof. Sandercock captures the spirit of this new paradigm by exhorting planning knowledge to expand to appreciate the significance of memory, spirituality and desire in the lived urban experience, all of which are crucially impacted by and can helpfully inform, a sensitive postmodernist planning practice.

Similar prescriptions for postmodernist planning were made by Dear (2000). He makes a call for pastiche and the importance of polyvocality. He looks to the drowning out of metanarrative in a plurality of claims and possibilities to inscribe the terrain of planning. Dear, like Beauregard (1989), argues that the form and function of the contemporary metropolis defies the logic of modernist planning, which was premised on stability and centrality. The polynuclear and flexible urban form of the postmodern city makes such master narratives irrelevant. Soja (1997) goes as far as to say that the diverse spatialities of contemporary capitalism mean that we are in fact 'postmetropolis'.

What Dear (2000) points up is the frequent impossibility, and indeed undesirability, of connected discourse. The postmodern metropolis (of which Los Angeles is the exemplar, he argues) is a pastiche of difference. Planning should not seek to cross discursive realms and should avoid pretensions of meta-theory to which, in its modernist form, it mistakenly made claim. This means a much more modest role for planning, or possibly even its demise (Dear 2000: 301). What the influence of postmodernism, in some small way, has done for planning is to problematize the idea of space.

I want to take up these challenges offered by postmodern planning in two ways. Whilst agreeing with its non-foundationalism and emphasis on difference, I argue that this does not entail a rejection of rationality or of dialogue across difference. Secondly, I agree that postmodernism has problematized space and suggest how a transactional rationality too accompanies a changed understanding of urban space and the role of planning within it.

PRAGMATIST PLANNING AND RATIONALITY 2: TRANSACTIONAL RATIONALITY AND TRANSACTIONAL SPACE

There are in fact a number of similarities between Beauregard, Sandercock and Dear's visions of postmodern planning and the deliberative paradigm of Forester and Healey. All call for the greatest possible range of voices and positions to be encompassed within the planning process. Neither assume that subjectivities are necessarily stable. All acknowledge the significance of situated knowledges. The importance of communicative rationality is in fact conceded as an important component of any postmodern planning effort by Sandercock: 'we also need a different kind of rationality that focuses on the formulation of goals. Rather than being technically based, this is a more communicative rationality with greater and more explicit reliance on practical wisdom' (Sandercock 1998: 205).

Harper and Stein (1996) ask the question whether the eclipse of the rational model and the uptake of pragmatist philosophy in planning are in fact related. They critique postmodernist planning on a number of grounds. First the assumption that language games are incommensurable in fact assumes a transcendent position of adjudication in order to suggest this in the first place, and foundational and transcendent positions are what postmodernists are trying to undermine (Harper and Stein 1996). They argue further that these absolutist assumptions are evident when postmodernists assume that the removal of foundations means the collapse of the whole system into an endless play of difference. It is possible, argue Harper and Stein, to have an idea of rationality and the justification of claims that is not foundationalist.[1] Harper and Stein's neopragmatist alternative – that no foundations does not mean incommensurability – sees difference as dialogue not incommensurability. Their idea of rationality draws heavily on the critical version of rationality that Habermas is advancing.

Niraj Verma (1996, 1998) also has an interpretation of rationality from a pragmatic perspective. He contrasts systemic reasoning with analytical reasoning. Whereas the latter isolates singular connections as causalities and looks for differences between phenomena, systemic reasoning looks for similarities between ideas and looks to purposes and ends rather than causes and mechanisms. Inspired by the work of William James this pragmatic rationality looks at the relations between phenomena as ones that exclude other potentialities. It is a way of looking at purposive connections rather than essential or logical connections. It tolerates indeterminacy, indeed finds that essential in the ongoing process of understanding. Verma develops James's 'sentiment of rationality' which is a passion to get acquainted with particulars. Verma's version of rationality criticizes analytic reasoning because it reduces its subject matter to a few issues while simultaneously marginalizing all others. What Verma is doing potentially is introducing a systematic form of reasoning into the professions (including the planning profession) that will accord with the inclusionary prospectus from deliberative planning. It is also one that takes in the emotions and the importance of context – elements again excluded from the traditional operation of rationality.

What Harper and Stein and Verma are pointing towards is a rationality that finds connection through difference. What I am suggesting for planning is a rationality based not upon prediction and prescription (as in the rational comprehensive model), nor deliberation and consensus (as in the Habermasian model) but rather a rationality built out of argumentation and dissensus.

PLANNING AS ARGUMENTATION

Traditional planning theory (in the rational comprehensive mode) sees planning as the ability to anticipate the future or resolve current conflicts. These require a precise definition of the objects (land uses, building

---

[1] Habermas's communicative rationality is not substantively foundational but its procedures are supported by an assumption of the universal pragmatics of language, which could be argued to be foundational.

development) and the interests in play. Resolution is made on the basis of technical specifications of efficiency in an instrumental (means–ends) logic. In deliberative or collaborative planning the emphasis is on the planner being outside of the planning department of city hall and immersed in the communities of interests related to the planning issue at hand. Deliberative planners stress that part of the planner's job that is about negotiation and the appreciation of other's worldviews and their claims. The purpose of planning is to bring the different validity claims into play in order to examine their good reasons or grounds through deliberative debate in order to come to an informed consensus.

Critics of deliberative planning point to its naivety about structures of power (see Huxley 2000; Huxley and Yiftachel 2000), or the fact that by privileging a certain form of linguistic interaction and style of debate, it excludes difference (Fraser 1992; Young 1990). Planning as argumentation addresses both these complaints. It supports difference in two ways. First it does not assume that participants come primed and ready to defend validity claims by rational criteria deployed in spoken discussion. To be fair to Habermas he is not assuming that participants are experts in debate in this sense. He believes that everyday intimate, or mundane, communication rests on the same validity redeeming criteria as public debate. Nevertheless these skills are implicit for the most part and it is reasonable to assume that participants would require some adjustment to be able to realize these social competences in this explicit way. Participants in argumentation are not assumed to be fully fledged debaters. Indeed, Hicks and Langsdorf (1999) argue that the process of participating might be constitutive of participants who are able to deliberate. Furthermore the sort of communication in play is not limited to linguistic action as it is for Habermas. Gestures, feelings and other performative aspects of communication come into play. These can be significant in a number of ways. Mutual attention to these communicative signals increases sensitivity to the other which can help in subsequent communication and the resolution of claims. Sometimes too they can communicate what participants really mean, aside from what they are actually saying, a point that Forester in a deliberative context has realized for some time (see Forester 1989, 2000). Furthermore non-verbalised and bodily expressions can be particularly fruitful in 'presenting diverse and unrepresented (indeed, as yet unverbalisable) goals, including those that Habermas excludes as strategic rather than communicative' (Langsdorf 2002: 147). This might be of particular importance for new or traditionally disempowered participants.

As I suggested earlier in considering the public realm (Chapter 5) Langsdorf adapts Willard's (1989) theory of argumentation that sees it as a communicative rather than logical activity. It can also be seen as a way of expanding enquiry by exploring the dimensions of oppositions, rather than the more usual assumptions about it bringing enquiry to a close. Also 'interpretation is a crucial activity in everyday conversational argumentation that occurs in plurivocal natural language, as distinct from formal argumentation that occurs in a normed, univocal language' (Langsdorf 1997: 315). Interpretation is an ongoing competence in everyday life. Participants must construe each situation they face and that construal is in part constitutive of that situation (including the heavy influence of habit and disciplinary mechanisms). Construing a situation involves linguistic and non-linguistic competences. Thus whereas for Habermas the function of linguistic action is to correctly represent participants' objective social and subjective views of the world and to approximate some idea of truth through deliberation, constitutive communication focuses on speech and the ethno-conventional negotiation of claims. Willard (1989) sees argumentation as conversational interaction based on dissensus. Locating argument in everyday social spaces means that a diversity of claims and dissensus is at the heart of this understanding. This contrasts with argument as logical progression in that the latter requires the progressive narrowing down of argument to truth or agreement. On the other hand it does not result in an endless play of difference as the postmodernists claim. The approach 'retains both a ground (dissensus) and a goal (expanding inquiry) for valorising conversation that thrives on controversy' (Langsdorf 1997: 316). Whilst not aiming at some singular truth, different positions might come to agreement about the plausibility of these positions. These positions do not represent subjective ideas or some objective correspondence with nature but rather are 'public' positions within transactions in an emergent intellectual ecology. This ecology is not seeking to limit arguments according to certain transcendent rules but is open to innovation, as Willard puts it 'conceptual changes . . . are likely adaptions blown in from the periphery. Plausibly, such innovations are made possible by the creation of new contexts, new connections among constructs, rather than conventional adaptions of existing constructs' (Willard 1989: 268).

The goal, Langsdorf argues, is to devise means to accommodate diverse positions rather than reduce them to one argument that wins. This links to the idea of 'rationalities . . . pragmatically linked to communities and contexts' (Willard 1989: 158). Accommodating difference in this way 'may be the valorisation of local agreement and recognition of multiple appropriate responses, the inevitability of negotiation, and the persistence of change' (Langsdorf 1997: 320). Argumentation is then 'what people do to create change' (Willard 1989: 262) rather than coming to some universal agreement (Langsdorf 1997: 320).

This situated idea of argumentation also brings with it some of the exclusions of 'propangandistic regimes' of disciplinary discourse (Willard 1989: 234) that are a block to openness. Within this situated and difference-oriented idea of argumentation are also the agonistic elements within discourse itself. There is the agon within communities and the agon between ideology or discourse (the distortions of power – what Willard calls propagandistic regimes) and alternative voices. Agonistic relations go all the way down. It is unlikely that there is some preserve of the lifeworld that permits non-instrumentalized language free from the pressures of the system (as Habermas hopes). All elements of life are subject to the distorting powers of ideology and resistance to this consists not in some safe haven (such as the community, or the lifeworld) for the reconstruction of meaningful communication. What is required is 'persistent argumentative practice . . . to preserve uncertainty despite the appeal of complacency' (Langsdorf 1997: 324). Thus although communication here is on the basis of situation-specific socio-spatial contexts it does not ignore wider circulations of power as discourse (that critiques of the communicative turn have often accused it of: Huxley 2000; Huxley and Yiftachel 2000).

This dissensus view of communication sees agonistic relations not just between communities, but within communities and even within discourses and voices. The role of rationality here is the attempt to accommodate diverse impulses and dissensus between communities, within communities and within the self. As Dewey argued growth of the self involves an ever more sympathetic understanding of diverse interests and impulses (Rosenthal 2002). This is transactional rationality at work in the widest sphere of communicative action. It is the clarification of competing feelings and interests. In term of argumentation it involves appreciating the plausibility of diverse positions and their inner connections to disparate situations. These inner connections are interpretative devices that draw on the situation as well as durable or institutional discourses, or extant meanings in social time and space. Interpretation goes on in the conversational gambits of everyday life and looking at the connective tissue of interpretation across disparate situations is the first step in assessing the plausibility of positions. This assessment is made on the move, between social spatial contexts, rather than with regard to externalities of objective relations or internalities of psychological states (Langsdorf 1997).

Planning as argumentation is an insurgent practice. The negotiation of claims is ethno-methodological (rather than logical) and so must be sensitive to the multiplicities of urban space and conceptions of space. There is no spatial common ground here, as is assumed on the isotropic plain of Cartesian rationality. Nor is there an emerging singular space of the public realm achieved via deliberation to consensus (in the Habermasian model). What we are left with are multiple spaces and multiple conceptions of space that transact as positions in argumentation.

Any given arena of planning argumentation will involve conflicting spatio-temporalities that are not confined to community. It involves an acknowledgement that interests in the situation are not necessarily clearly formed in all cases, and it implies a critical sensitivity to powerful discourses. It suggests a planning imagination that does not default to the most technically efficient solution, nor to an easy idea of consensus. Using a transactional rationality the pragmatist planner searches out dissensus, unformulated expression and the critical relation of the planning situation to other spatio-temporal realms. This is not an endless openness as the postmodern planners suggest. It seeks connection through diversity and difference and it is attendant to the situational nature of time and space.

8 Cosmopolitan reason and the global city

In 'The Village of the Liberal Managerial Class' Bruce Robbins (2001) makes a comparison between two Booker-prize-winning novels – Ishiguro's *The Remains of the Day* and Ondaatje's *The English Patient*, both of which feature cosmopolitans who are also aristocrats. In *The Remains of the Day* Lord Darlington out of loyalty to German fellow aristocrats cuts across national identity to produce a fatal complicity with Nazism. In Ondaatje's novel the aristocrat cosmopolite is Lord Almasy, the so-called English patient, who experiences a dream of the desert and a love affair that breaches national borders but is shattered by nationalisms and exclusions of the Second World War. Robbins links these treatments to discussions of cosmopolitanism defined as a kind of expertise and rationality, a framework of skills and knowledge that translate anywhere: professionalism as a new form of world aristocracy.

What unties all the main characters in *The English Patient*, Robbins argues, is the love of knowledge. The metaphor of education as exploration is developed through the story of an exploration in the desert, where the boundaries between Europe and Africa have become blurred, but also where there is a freedom from nation, and education and knowledge as erotic adventure. Thus 'Almasy's erotic bond with Katherine Clifton *has the same content* as his homosocial bond with Madox and the other men: an eroticising of cosmopolitan knowledge' (Robbins 2001: 24; emphasis in original). *The English Patient* suggests for Robbins a 'postpatriotic love' (Robbins 2001: 26).

In *The Remains of the Day* love and professionalism are set against each other. Stevens, the butler, sacrifices love with Miss Kenton for professional duty to his employer. It also extends to his moral failure in backing his employer's support for the Nazis. Lord Darlington convenes a conference of diplomats in 1923 to attempt to ease German reparations. In the following years his anti-Semitism and attempts to stop the war align him with the Nazis. Stevens' moral and personal failure is emphasized by his continuing with his professional duties in serving the diplomats as his father lies dying upstairs. Told of his father's death he responds by saying he is 'very busy just now'.

Robbins points out that the story is complicated by the fact that, in terms of international relations, easing of reparations might have indeed prevented the rise of the Nazis and the Second World War, as notable figures such as Maynard Keynes were arguing at about the same time as Darlington's fictional conference. The character of Lewis the American senator is revealed to be duplicitously undercutting Lord Darlington to ensure reparations to America. Lewis rises to his feet at one of the conference dinners to tell his host and fellow guests 'international affairs today are not longer for gentlemen amateurs'. He then offers a toast 'to professionalism' and no one drinks but himself.

Robbins offers a more positive account of professionalism against this negative portrayal in *The Remains of the Day*. From the story itself there is the character of Harry Smith, as his name suggests representing the common man, who defends the war against the Nazis and argues that ordinary people can have strong feelings on all kinds of things. Yet this democrat at home is also an imperialist abroad, asserting that all the colonies should not be allowed independence. Furthermore, Robbins suggests, the butler's failure to realize his love for Miss Kenton was not because of the professional constraints of his work but that the affection for her resulted from their shared love of professional competence. This is the eroticization of knowledge through professional conduct that is evident in both *The English Patient* and in Robbins' sense of it, in *The Remains of the Day*.

It is not just the specialized knowledge that is significant in the understanding of cosmopolitanism, but the way that it is deployed. As the example of the butler in *The Remains of the Day* suggests, it involves the learning and enacting of certain ritual performances. And also in the case of Stevens it entailed a commitment to procedure at the expense of local commitments (witnessed in his obtuseness and distance from those close to him). For the explorers in *The English Patient* their recognition of the love of knowledge was the commonality that cut across differences of gender and ethnicity. This love of knowledge might also be thought of as cold and calculating, although as Robbins' article points out, it might not be a rejection *of* love but a form of love that is beyond local loyalties.

These are the characteristics of cosmopolitan professionalism: acquisition of relevant specialist knowledge and its ritual deployment, free from local passions and in form of conduct that is transferable between very different situations. These are the assumptions behind the growing literature on the associations between professionalism, rationality and the rise of a global new middle class (Gouldner 1979; Ehrenreich and Ehrenreich 1979; Isin 2000; Hannerz 1996). The association between ideas of cosmopolitanism and the rise of a new middle class has been established at least since 1979 when Alvin Gouldner argued that the defining characteristic of the new class was its cosmopolitanism. Hannerz (1990) develops the distinction between locals and cosmopolitans with the influence of the former being based not on what they know but who they know, whereas for cosmopolitans the relationship is reversed.

> Cosmopolitans . . . base[d] whatever influence they had on a knowledge less tied to particular others, or to the unique community setting. They came equipped with special knowledge, and they could leave and take it with them without devaluing it . . . they are the 'new class', people with credentials, decontextualised cultural capital.
>
> (Hannerz 1996: 108)

This knowledge is accompanied by a set of analytics – 'a culture of critical discourse' (Gouldner 1979; Hannerz 1990).

A contrasting approach to the discussion of professionalism, rationality and the new middle class is that cosmopolites demonstrate an openness to difference, that the deployment of their rationality is marked by its sensitivity to difference. Recent writing on cosmopolitanism (Hannerz 1996; Dharwadkar 2001; themed issue of *Theory Culture and Society* 2002 vol. 19: part 1/2) seem to stress it ideally being defined by mobility and a sense of openness to otherness.

> Of importance for cosmopolites is openness to the other, as Gilles Deleuze repeatedly puts it, to think towards the horizon; to alter one's perceptions and to invent new 'smooth spaces' instead of building walls, barriers and prisons to conserve one's interests.
>
> (Conley 2002: 131)

Conley (2002) ties this to the type of rhizomatic connections that Delueze and Guattari postulate. To be a cosmopolite is to occupy a junction in a rhizomatic network where there are only lines of flight to other diverse connections. These connections are themselves only temporary and partial in a way that Deleuze defined as choasmopolis (cited in Conley 2002: 131).

Dominant contemporary manifestations of cosmopolitanism involve knowledge based on professional competences and transnational connections. The degree to which this is a movement towards a global democracy, or a new transnational exclusivity of elites, is at the core of these debates, as are questions about whether the openness to the other is genuine or a kind of fake, consumption-oriented transnational flaneurism (May 1996; Hage 1997). Also central are questions about the degree to which cosmopolitanism is or is not tied up with cities, especially global cities. It is a form of knowledge and conduct that, in its concentrations and connections, might be associated with certain urban spaces, especially particular districts of global cities (Sassen 2000).

Elite enclaves in global cities are very much part of the infrastructure of what Castells (1996) calls 'the space of flows'. The space of flows is 'the material organisation of time-sharing social practices that work through flows' (Castells 1996: 412) where flows are defined as 'purposeful, repetitive, programmable sequences of exchange and interaction between physically disjointed positions held by social actors in the economic, political, and symbolic structures of society' (Castells 1996: 412). The space of flows is supported by circuits of electronic impulses (microelectronics and telecommunications and so on); by nodes and hubs in these networks, of which the 'global city' is a key instance; and by the spatial organization of the transnational managerial elite. The spatial organization of the dominant elites form landscapes of power (Zukin 1991) that provide a seamless environment for work and leisure, all over the world. It involves a symbolic economy that excludes the masses by the enormous amounts of economic and cultural capital required for membership. The symbolic economy is a manifestation of 'good taste' in all aspects of life, from body presentation and deportment, through exercise, diet, desirable residential address, home furnishings – in a set of aesthetic dispositions that are increasingly transnational. Some aspects of this are captured by my work on the gentrification premium (Bridge 2001a, b, discussed in Chapter 6). These elements of lifestyle symbolic capital are supported when cosmopolites are on the move, via airport VIP lounges, exclusive hotels with familiar ambiances and technical and personnel support to facilitate constant communication and a comfortable working environment.

The rationality at work in the space of flows is consistent with the nature of the flows themselves: 'purposeful, repetitive and programmable'. It is the abstract, technical rationality guided by instrumental means to technically efficient ends. It is expert knowledge that claims universality in being applicable, or adaptable throughout the world. Its socio-spatial manifestation is a form of communicative rationality presaged on a desire for status and security in the face of difference (which I explored in the discussion of elite gentrifiers in Chapter 6).

A further aspect of the knowledge of the professional managerial class, originally defined by Ehrenreich and Ehrenreich (1979), was the management of the reproduction of capitalist social relations. This culminates, in Castells' terms, as the segmentation and disorganization of the masses. The masses themselves occupy the 'space of place', 'a locale whose form, function and meaning are self-contained within the boundaries of physical contiguity' (Castells 1996: 423). The space of place can have positive attributes, in forms of community identity or the reassuring familiarity of a neighborhood. But it can also promote insularity, a limited outlook and disconnection from wider power relations. The form of rationality most associated with the space of place is practical reason. This is the form of common sense, based on local history and social norms, that is limited to community or class (Bourdieu 1977, 1984).

It is possible to think of the global city as consisting of two spaces of rationality: cosmopolitan rationality (increasingly bound up with the knowledge requirements of global capitalism) operating in a space of flows, and 'local' rationality existing in a plurality of different socio-cultural community contexts defined by the space of place. Sassen (2000) sees the financial districts of the global cities of New York, London and Tokyo as city-states: vitally connected to each other across the globe and, at the same time, seemingly disconnected – economically, intellectually and morally – from the rest of the city that surrounds them. Cosmopolitan rationality is detached by its specialized techniques from local concerns just as in *The Remains of the Day* the butler focuses on his ritual duties to the exclusion of his local relations and emotions.

But what of the version of cosmopolitan reason projected by the protagonists of *The English Patient* – cosmopolitan knowledge as a sort of postpatriotic love? The relations between the characters cut across difference but they are equivalent in that the particularities of their relationships are sublimated by their love of knowledge. The eroticization of knowledge is the sense of adventure and exploration in the desert. The desert is an extraordinary space, the meeting of Africa and Europe, exotic and unexplored, and in these senses separate from the ordinary spaces of everyday life. Cosmopolitan reason is again an elite knowledge at work in a space that allows freedom of exploration. Detachment from local ties is in a form of ritual in *The Remains of the Day* and mobility in *The English Patient*.

Rationality as ritual and mobility is to some extent captured by the idea of transversal rationality. Welsch (1998) identifies a form of cosmopolitan reason as transversal reason, a pure faculty that is based on the 'logic of transition', or ability to move between rational spheres. This contrasts with traditional rationality, which was a logic of identity. Schrag's (1992) idea of transversal rationality also resonates with this sense of cosmopolitanism. It arises out of community context but involves the capacity to be distant from community norms. Transversality avoids the verticality or hierarchy of traditional rationality as the logic of transition, and the horizontality of an endless play of difference. It cuts across difference obliquely. Schrag calls on Bakhtin's idea of chronotope as the socio-practical constitution of time and space through dialogue. 'Rationality is the ability to discern and articulate how our discursive and non-discursive practices hang together in the texture of everyday life' (Schrag 1992: 94). In the relation with 'the other', that key moment of cosmopolitanism, there is claim to 'convergence without co-incidence' (Schrag 1992: 158).

In contrast to transversalism, the idea of transactional rationality that I have pursued throughout this book involves community knowledge but then the encounter with the other as particularity. This cosmopolitan idea of community participation and distance parallels Dewey and Mead's conceptions of the tension between the 'I' and the 'me', the creative historical subject and the role-taking member of a community. And here also are the two critical moments of transactional rationality. The Meadian 'me' fosters rationality as the anticipation of the reaction of the generalized other, or the entire community. That anticipation hinges on the operation of shared symbols, most notably in language, that call forth the anticipated response. Unanticipated and non-discursive actions are constitutive of communication, but call forth a response to the *particular* other in ways that are constitutive of the situation and rely on novel responses. The ability to improvise and to respond to particularity is where role-playing rationality gives way to situational reasonableness. The 'I' is the unpredictable historical agent, the improviser. The 'I' is both affirmed and made uncertain through the absolute alterity of otherness, in the form of subjectivities or objects (all of which, as we have learned from Dewey, are forms of transaction).

Drawing on the orientations from Dewey these communicative spaces are transactional and processual. As well as discursive communication they involve the whole panoply of communicative activities – chronemics, haptics, kinesics, proxemics and vocalics – the ways that everyday performing bodies interact, that defy representation (Langsdorf 2002: 147). Such spaces of communication are transactional but not always co-present. Even where co-present they are often overlooked.

It is in the city, that being together of strangers, that the multiplicity of spaces of communication still has the greatest potential. Whatever the levels of transaction there is at least a sight of difference in the city. The city also provides the background latent capacities against which communicative action can be realized. Counter hegemonic discourses, or subaltern counter publics, are more possible against the background communicative noise of the city.

These spaces of communication both reproduce larger power discourses but also use strategic competences to their own advantage. They exist within the system rather than being separate from the system. And they are often overlooked because the content of communication is tacit and their potential for change is latent. The transformation from latent to active engagement can be seen in reaction to trauma. Witness the personalized grieving of New Yorkers assembling in crowds and holding up photographs of lost loved ones in the wake of 9/11. In March 2004 there were the scenes of collective silent vigil of eight million Spanish residents in the streets of various cities in response to the 11 March bombings of the trains in Madrid. These are traumatic instances of the communicative latency of the city, but we have also explored more positive instances such as the crowds that gather to watch Sydney Mardi Gras (see Chapter 3).

The examples discussed are very marked instances of collections of bodies, emotions and minds on the streets of the city but the latent potential of the city is built out of communicative exchanges of minute ritual courtesies that are the basis of civitas – of city life. These communicative exchanges have been largely overlooked by the focus on symbolic interaction at the expense of non-symbolic interaction and because instances of minor cooperation are less exciting to report (Lofland 1998). Such situations are often more difficult to research because they are not demonstrative and conspicuous.

THE NEIGHBOURHOOD AND DIFFERENCE

I suggest that ongoing, everyday transactional rationality is part of the life of the urban neighbourhood. MacCannel (1999) critiques the New Urbanism movement in the USA through a comparison with a more 'ordinary' urban neighbourhoods. New urbanism or neotraditionalism is the movement to build high-density, middle class suburban developments or urban infill that are designed to resemble a nineteenth century town. Despite their claims to give a renewed sense of community and neighbourliness, MacCannel argues that they are in fact built out of 'statistical' ties of equivalent social standing and social homogeneity, rather than any deeper community loyalties. New Urbanism reaches its high tide mark in the town of Celebration in Florida designed to be the perfect community by the Disney Corporation. Here all aspects of community life are legislated for and the 'panoptical' houses have clear lines of sights through each dwelling to permit the surveillance from outside. New Urbanism neighbourhoods are based on an explicit and abstract rational contract over neighbourhood conduct. In that sense they are equivalent to the written contracts that control common interest developments, or gated communities (McKenzie 1996).

MacCannell contrasts the neighbourhoods of New Urbanism with the seemingly drab anonymity of the more run-down neighbourhoods that are revealed by the reality TV show *Cops*. Following the activities of the cops on their raids, or giving assistance to domestic problems, is breaking the rule against sudden incursion deep into domestic space. What these visits do, according to MacCannell, is reveal the differences that exist within what initially appears as a drably homogeneous urban space. A pursuit through a series of back gardens, and the minute details of domestic arrangements, reveal an unanticipated degree of variation in lives. MacCannell also notes the ability of people to normalize neighbours, however different. Called to a disturbance at a house, police find a wild pig on the premises. In questioning the neighbours they discover that the pig is often at large in the neighbourhood, accompanied by a dog, both treated by those residents questioned as occasionally annoying neighbours.

For MacCannel the *Cops* neighbourhoods are symbolically rather than statistically mediated.

> Not every house or apartment in *Cops* is 'the same' as the one beside it, but they are ineffably equivalent in the moral totalisation of the 'neighborhoods'. This moral homogeneity appears to be based on contractual and quasi-contractual agreements. This is not just a matter of demographics, of lending policies, zoning laws, residential inspection practices and the like. And it is certainly not a matter of centralised rules governing the color of curtains, or parking for pick-up trucks. Local tolerance of disorder and noise levels, the appearance of front yards, behaviour of children and dogs, even pigs, of what are taken to be 'appropriate' ways

of appearing to be different, of the local limits of difference; all of this must be worked out between neighbors as a living agreement. It is precisely the capacity to arrive at this kind of agreement with one's neighbors which has been denied by the New Urbanism.

(MacCannell 1999: 120)

MacCannell argues that 'the enormous variation of neighborhoods in America are the laboratories in which new cultural arrangements are being created by people . . . culture guarantees human intelligence by not offering any guarantees on human interaction' (MacCannell 1999: 120). We saw in chapter 4 from Chauncey's work how proximity of different communities and the non-discursive and discursive interactions helped forge the range of gay identities in the emergence of the gay community in New York.

Neighbourliness involves elements of closeness and distance and neighbourly relations are a good example of an ongoing transactional rationality that is both discursive and non-discursive. Neighbourhoods are full of tacit communications that together make up the neighbourhood ambiance. They also involve particular tacit accommodations between neighbours. Tacit activities of this kind are full of communicative import as the game theorist Schelling noted.

If my neighbor's fruit tree overhangs my yard and I pick exactly all the fruit on my side of the line, my neighbor can probably discern what my 'proposal' is, and has a good idea of what he has acquiesced in for the future if he does not retaliate. But if, instead, I pick that same amount of fruit from both sides of the line haphazardly or pick some amount that is related, say, to the size of my family, he is less likely to perceive just what I have in mind. (He may also be more obliged to resist or retaliate if I pick only *part* of the fruit on my side of the line than if I pick it all, since I have failed to demarcate the limit of my intentions).

(Schelling 1980: 104, fn)

Neighbourly relations are full of these discursive and non-discursive rational negotiations.

Much of neighbourhood interaction is tacit, or at the edges of discursive communication, what de Certeau (1984) called the 'phatic' aspect (see Chapter 3). Neighbourly interaction has that quality of constantly settling the terms of interaction through rituals of greeting and signals of the degree of openness to further interaction. Neighbours engage on a daily basis in a range of discursive and non-discursive rituals that communicate the degree to which they are 'available' for further communication. An incline of the head, or tap of the watch, or facial gesture of exhaustion, signal that the participant wants to keep walking. A check in the step, the raising of eyebrows, a more open body aspect can suggest availability for conversation. These ritual and daily negotiations can have greater significance in thinking about communication in public. As Young (2000) has noted, rituals of greeting are critical in setting the tone and conditions for fuller communication in public (which I discussed fully in Chapter 5).

TRANSACTIONAL RATIONALITY AND COSMOPOLITAN URBAN SPACE

Dewey had reservations about the growth of a technical professional class. He was sceptical of the role of experts in democratic communities: 'as if there is a group of individuals by virtue of their "scientific" expertise better able to make the types of judgements and decisions required in our everyday lives' (Bernstein 1998: 149). 'A class of experts is inevitably so removed from common interests as to become a private class with private interests and private knowledge' (Dewey 1927: 148–9). The distanced professionalism of the butler in *The Remains of the Day*, is mirrored by the social and spatial separation of the professional/managerial elite in the global city.

In contrast to expert knowledge is Dewey's idea of social intelligence, which in its highest form is based on openness, fallibility, experimentation, ongoing criticism and imagination of whole community. That comes out of the diversity of communication in different socio-cultural contexts. This seems to sit well with an idea of transversal rationality, a logic of transition, a circulatory rationality that is able to move between rational spheres whilst still being articulated from socio-cultural difference. It also resonates with certain claims for postmodern urban spaces. These see the city as surface, decentred, with no overall logic, a play of difference, polyvocal, a pastiche (Dear 2000; Gibson and Watson 1995; Soja 1989). The city can be seen as an assemblage of effects (human and non-human), a place of 'present becoming' (Deleuze and Guattari 1998). Space is relational and horizontal, disrupting hierarchy, and denying fixed meaning as an interplay of diverse discourses.

The 'Deweyan' city is indeed processual, emergent and radically perspectival. Yet there are, I argue, certain orderings still at work. They are not the singular orderings of one rationality, or one public realm, or even *the* city. There are the orderings of power discourse that imbue actors with certain habits of action or dispositions towards action. These disciplinary spaces of the city have been the subject of acute analysis and critique by urbanists (Davis 1990; Harvey 2000; Mitchell 2003). Yet within and beyond these disciplinary spaces are the myriad space–times of communication between and within socio-cultural contexts. They are often mundane, sometimes imperceptible, often overlooked, but they represent a kind of background hum of the city. This comprises socio-spatial realms of communication, of varying intensities and durations, which overlap and are linked to other communicative realms in a complex network of communicative action. These are not formless and ever-becoming but involve conflict and dissensus within and between discursive communities. Through forms of argumentation and interpretation, mutual claims are examined and their validity assessed. There can be non-communication and the silence of social division. But as we saw in Chapter 3 'On the street', even the most self-limiting forms of social interaction are leaky and suggestive of other communicative import.

These spheres of communication are constitutive of time and space. This is not the modernist city – a flat Cartesian plane linked by points and lines that result from the ordering of a singular instrumental rationality from above. Nor do we have the endless becoming, fragmentation, liminality and thirdness (Soja 1989, 1996) of postmodern urban spaces. Rather we have a myriad of time-space spheres of communication and articulation where claims are negotiated between socio-pragmatic situations (rather than appealing to a universal or singular public sphere). The city is more like a complex of bilaterals where the faculties of interpretation at work are themselves constitutive of time and space. And the reason that connection is possible across difference is that dissensus goes all the way down, into communities as well as between them, within voices and discourses as well as between them. It is also possible because interpretation goes on all the time, even in the most mundane spaces of everyday life. Extending the insights of the ethno-methodologists, all situations have to be construed in communication rather than being set time and space backdrops for action to commence.

Ethno-methodologies – the social competences and conventions that people use to get by in everyday interaction – nevertheless have ritual functions that might limit other-regarding communication. Often these conventions are inscribed on the body as communication, involving tacit knowledge (or class disposition in Bourdieu's terms). What we have learned in the course of the exploration of Dewey's thought (especially in Chapter 2) is that this body, tacit or 'had' knowledge is not necessarily a default position of communication to allow conventions to run smoothly, but is itself a form of ongoing intelligence, heavily influenced by prior habit but not confined to it. The other side to this is that situations are not always clear-cut – slippages and other infelicities in communication occur frequently. Rationality here is the re-establishment of some form of consistency of expectation via mutual self-control. But it is important to remember that problematic or unexpected situations can lead to communicative action that starts to lead in another direction, either because of the agency of the 'I', or the need to coordinate on a new 'me' or community norm.

As well as ongoing 'mundane' interactions there are a whole range of space–times of communication. There are long and thin reaches of instrumental or strategic action. Here space–times are ordered, put together in causal chains in which the means or the ways through these ordered situations are given emphasis. Pursuing strategic goals necessarily means that agents are more blinkered to the full range of communicative action that is available in each time–space situation they pass through. However, this closing down is never complete. Instrumental goals tend to be highly cognitive and so can be thrown off course by other cognitive agents (the stuff of strategic interaction and game theory). Strategic chains of time–space can also receive shocks via the intrusion of alterity (such as disease) or as a result of messages received from the ongoing sensory intelligence of transactional bodying. Projects can encounter occasions of situational excess (giving birth, bereavement, 'falling' in love) and these lead to the more intense and evaluative space–times of communicative action. And then there are the shocks to the system, or the moments of epiphany that have traditionally been represented through aesthetic communication, the extraordinary immediacy and vibrancy of 'world-disclosing' situations. The point is that these space–times are not individual experiences or via universal representations (such as neutral bodies or rational minds) but are situational transactions of body-minds.

This more horizontal and perspectival spatiality might seem to support ideas that subjectivity is now dispersed, or weak (Amin and Thrift 2002). Transversal rationality too follows a logic of displacement and transition, rather than identity and ego. As I see it the strength of subjectivity is strongly associated with the web of overlapping power networks in which the subject is located (Bridge 1997b; Bridge and Watson 2002). Rather than a weakening of subjectivity I suggest that it is just becoming distributed across more diverse networks. We have seen this in the idea of ambage (or circuitousness) from White's work on networks and developed by Emirbayer and Sheller (1999) in thinking about the public realm (which I discuss in Chapter 5).

It is here that I think there is a deepening as well as horizontalizing of space. There is another possible aspect of a rationality that stands under a multiplicity of perspectives, that it is a logic of transition of the self. If dissensus does indeed go all the way down, within voices as well as between them, that can be manifest in the competing thoughts, emotions and body inclinations in what we might call the self. Rationality for Dewey is involved in a reconstruction of the community and the self. For the self it involves 'an ongoing process of learning from and restructuring one's concrete self – in the fullness of its emotions, drives, desires, beliefs, purposes and aspirations – in the process of adjusting to the other' (Rosenthal 2002: 216). It implicates 'a vague sense of thick agency' (Rosenthal 2002: 213) that comes from the body as well as the mind (body-mind). It permits speech acts that are innovative, expressive of the 'I', rather than always being accompanied by the 'me' of community validation.

We can see this logic of transition of the self by returning to the idea of cosmopolitanism. For Dewey cosmopolitanism came out of situation rather than mobility or transition. Rationality is an aspect of the concrete human being in its entirety (Dewey 1922; Rosenthal 2002). It was about having to live with difference, rather than just moving between different contexts. In these conditions rationality was being fully lived. This was suggested in the work of Mary Parker Follett whose pioneering efforts in the social centres of the Progressive Era I considered in Chapter 5. Follett thought that the urban neighbourhood with its mixing of difference was the best example of a cosmopolitan space. This form of provincialism made local people more interesting than 'cosmopolitan people' who sought out homogeneous social situations (Mattson 1998: 96). Follett explained: 'Why are provincial people more interesting than cosmopolitan [people], that is, if provincial people have taken advantage of their opportunities? Because cosmopolitan people are all alike – that has been the aim of their existence and they have accomplished it' (Follett, 1918: 195, cited in Mattson 1998: 96). It was in the urban neighbourhood, not the gemeinschaft community or through transnational connections, that people had to face up to heterogeneity and where democratic spirit has the potential to emerge.

Truly cosmopolitan reason is not confined to elite knowledge or separate transnational spaces. It emerges in the urban neighbourhood in which difference is a daily reality and a negotiation at the very least at the 'phatic' edges of communication. This gives some prospect of a city that is not divided between cosmopolitan rationality of the global district and the practical rationality of the surrounding 'provincial city'. Rather it is a form of cosmopolitan rationality that comes from location, rather than transition, and that is beyond community. It is in the burgeoning transnational neighbourhoods of the global cities that the best prospects of cosmopolitanism are to be found: cosmopolitanism as a form of reason lived daily in the city of difference.

References

Aboulafia, M. (1999) A (neo) American in Paris: Bourdieu, Mead, and Pragmatism. In R. Shusterman (ed.) *Bourdieu: A Critical Reader*. Oxford: Blackwell.

Addams, J. (1968) *Twenty Years at Hull House*. New York: Macmillan (originally published in 1910).

Alonso, W. (1964) *Location and Land Use*. Cambridge, MA: Harvard University Press.

Amin, A. (2000) The economic base of contemporary cities. In G. Bridge and S. Watson (eds.), *A Companion to the City*. Oxford: Blackwell, 115–29.

Amin, A. and Thrift, N. (2002) *Cities: Reimagining the Urban*. Cambridge: Polity.

Anderson, E. (1990) *StreetWise: Race, Class and Change in an Urban Community*. Chicago: Chicago University Press.

Ansell Pearson, K. (1999) *Germinal Life*. London: Routledge.

Arendt, H. (1958) *The Human Condition*. Chicago: University of Chicago Press.

Arendt, H. (1968) Introduction. In W. Benjamin, *Illuminations* (trans. H. Zohn). New York: Harcourt, Brace and World.

Atkinson, R. and Bridge, G. (eds) (2004) *Gentrification in a Global Context: the New Urban Colonialism*. London: Routledge.

Austin, J. L. (1962) *How to Do Things with Words*. Oxford: Oxford University Press.

Bakhtin, M. (1984*)* *Rabelais and his World* (trans. H. Iswolsky). Bloomington: Indiana University Press.

Barnes, T. (1996) *Logics of Dislocation: Models, Metaphors, and Meanings of Economic Space*. New York: Guilford.

Bartlett, N. (1990) *Ready to Catch Him Should He Fall*. New York: Dutton.

Baudrillard, J. (1981) *For a Critique of the Political Economy of the Sign* (trans. C. Levin). St Louis: Telos Press.

Bauman, Z. (2000) *Liquid Modernity*. Cambridge: Polity.

Beauregard, R. (1989) Between modernity and postmodernity: the ambiguous position of US planning. *Environment and Planning D: Society and Space* 7: 381–95.

Beaver, H. (1981) Homosexual signs (in memory of Roland Barthes). *Critical Inquiry* 8.

Bell, D. (1976) *The Cultural Contradictions of Capitalism*. New York: Basic Books.

Benjamin, W. (1979) One way street. *In One Way Street and Other Writings*. London: Verso (originally published in 1925–26).

Benjamin, W. (1999) *The Arcades Project* (trans. H. Eiland and K McLaughlin). Cambridge, MA: Harvard University Press.

Bennett, A (1999) Subcultures or Neo-Tribes? Rethinking the relationship between youth, style and musical taste. *Sociology* 33: 599–617.

Bernstein, R. (1986) *Philosophical Profiles: Essays in a Pragmatic Mode*. Cambridge: Polity

Bernstein, R. (1998) Community in the Pragmatic Tradition. In M. Dickstein (ed.), *The Revival of Pragmatism: New Essays on Society, Law and Culture*. Durham, NC: Duke University Press, 141–56.

Blum, D. (1983) The Evils of Gentrification. *Newsweek* (3 January): 7.

Blumer, H. (1969) *Symbolic Interactionism: Perspective and Method*. Berkeley: University of California Press.

Bourdieu, P. (1977) *Outline of a Theory of Practice*. Cambridge: Cambridge University Press.

Bourdieu, P. (1984) *Distinction: A Social Critique of the Judgement of Taste*. London: Routledge and Kegan Paul.

Bourdieu, P. (1991) *Language and Symbolic Power* (trans. G. Raymond and M. Adamson). Cambridge, MA: Harvard University Press.

Boyer, C. (1983) *Dreaming the Rational City*. Cambridge, MA: MIT Press.

Bridge, G. (1994) Gentrification, class and residence: a reappraisal. *Environment and Planning D: Society and Space* 12: 31–51.

Bridge, G. (1995) The space for class: on class analysis in the study of gentrification. *Transactions of the Institute of British Geographers* NS 26: 236–47.

Bridge, G. (1997a) Towards a situated universalism: on strategic rationality and 'local theory'. *Environment and Planning D: Society and Space* 15: 633–9.

Bridge, G. (1997b) Mapping the terrain of time-space compression: power networks in everyday life. *Environment and Planning D: Society and Space* 15: 611–26.

Bridge, G. (2000) Rationality, ethics and space: on situated universalism and the self-interested acknowledgement of difference. *Environment and Planning D: Society and Space* 18: 519–35.

Bridge, G. (2001a) Bourdieu, rational action and the time-space strategy of gentrification. *Transactions of the Institute of British Geographers* NS 26: 205–16.

Bridge, G. (2001b) Estate agents as interpreters of economic and cultural capital: the gentrification premium in the Sydney housing market. *International Journal of Urban and Regional Research* 25: 87–101.

Bridge, G. (2003) Time-space trajectories in provincial gentrification. *Urban Studies* 40: 2545–56.

Bridge, G. and Dowling, R. (2001) Microgeographies of retailing and gentrification. *Australian Geographer* 32: 93–107.

Bridge, G. and Watson, S. (2002) Lest power be forgotten: networks, division and difference in the city. *The Sociological Review* 50: 505–24.

Bruce, I., Murphy, P. and Watson, S. (1997) Gay sites and the pink dollar. In P. Murphy and S. Watson (eds). *Surface City: Sydney at the Millennium*. Sydney: Pluto.

Buchholz, R. and Rosenthal, S. (2000) *Rethinking Business Ethics: A Pragmatic Approach*. Oxford: Oxford University Press.

Buck-Morss, S. (1989) *The Dialectics of Seeing: Walter Benjamin and the Arcades Project*. Cambridge, MA: MIT Press.

Burgess, E. W. (1925) The growth of the city: an introduction to a research project. In R. E. Park, E. W. Burgess and R. D. McKenzie (eds). *The City*. Chicago, IL: University of Chicago Press.

Butler, J. (1993) *Bodies That Matter*. London: Routledge.

Butler, J. (1997) *Excitable Speech: A Politics of the Performative*. London: Routledge.

Butler, T. (1997) *Gentrification and the Middle Classes*. Aldershot: Ashgate.

Butler, T. and Robson, G. (2001) Social capital, gentrification, and neighbourhood change in London: a comparison of three south London neighbourhoods. *Urban Studies* 38: 2145–62.

Callon, M. (1991) Techno-economic networks and irreversibility. In J. Law (ed.), *A Sociology of Monsters: Essays on Power, Technology and Domination*. London: Routledge, 196–261.

Castells, M. (1977) *The Urban Question*. London: Edward Arnold.

Castells, M. (1983) *The City and the Grassroots: A Cross Cultural Theory of Urban Social Movements*. London: Arnold.

Castells, M. (1996) *The Rise of Network Society, vol. 1: The Information Age: Economy, Society and Culture* (3 vols). Oxford: Blackwell.

Castells, M. (1997) *The Power of Identity, vol 2: The Information Age: Economy, Society and Culture* (3 vols). Oxford: Blackwell.

Castells, M. (1998) *End of Millennium, vol 3: The Information Age: Economy, Society and Culture* (3 vols). Oxford: Blackwell.

Chauncey, G. (1994) *Gay New York: Gender, Urban Culture and the Making of the Gay Male World 1890–1940*. New York: Basic Books.

Chtcheglov, I. (1958) Formulary for a New Urbanism. *Internationale Socialiste* 6 August 1961, and *Situationist International Anthology*, p. 3.

Clay, P. (1979) *Neighbourhood Renewal: Trends and Strategies*. Lexington, MA: Lexington Books.

Cohen, M., March, J. and Olson, J. (1988) A garbage can model of organizational choice. In J. March (ed.), *Organizations and Decisions*. New York: Basil Blackwell.

Collett, P. and Marsh, P. (1974) Patterns of public behaviour: collision avoidance on a pedestrian passing. *Semiotica* 12: 281–99.

Collins, R. (1981) On the microfoundations of macrosociology. *American Journal of Sociology* 86: 984–1014.

Conley, V. A. (2002) Chaosmopolis. *Theory, Culture and Society* 19: 127–38.

Crossley, N. (2001) *The Social Body*. London: Sage

Davis, M. (1990) *City of Quartz: Excavating the Future in Los Angeles*. London: Verso.

Dear, M. (2000) *The Postmodern Urban Condition*. Oxford: Blackwell.

Debord, G. (1957) Report on the Construction of Situations and on the International Situationist Tendency's Conditions of Organisation and Action. *Situationist International Anthology*.

De Certeau, M. (1984) *The Practice of Everyday Life*. London: University of California Press.

Deleuze, G. and Guattari, F. (1988) *Thousand Plateaus: Capitalism and Schizophrenia*. London: Athlone.

Deutsche, R. (1999) Reasonable urbanism. In J. Copjec and M. Sorkin (eds), *Giving Ground: The Politics of Propinquity*. London: Verso, 175–206.

Dewey, J. (1922) *Human Nature and Conduct*. New York: Henry Holt & Co.

Dewey, J. (1927) *The Public and Its Problems* New York: Henry Holt & Co.

Dewey, J. (1938) *Experience and Education*. New York: Macmillan.

Dewey, J. (1958) *Experience and Nature*. New York: Dover (originally published in 1929).

Dewey, J. (1987) Art as Experience. In J. A. Boydson (ed.), *Later Works*, Vol. 10. Carbondale and Edwardsville: Southern Illinois University Press (originally published in 1934).

Dewey, J. (1991) Logic: The Theory of Enquiry. In J. A. Boydson (ed.), *Later Works*, Vol. 12. Carbondale and Edwardsville: Southern Illinois University Press (originally published in 1938).

Dewey, J., and Bentley, A. (1991) Knowing and the Known. In J. A. Boydson (ed.), *Later Works*, Vol 16. Carbondale and Edwardsville: Southern Illinois University Press (originally published in 1949).

Dharwadkar, V. (2001) *Cosmopolitan Geographies: New Locations in Literature and Culture*. London: Routledge.

Dickstein, M. (ed.) (1998) *The Revival of Pragmatism: New Essays on Social Thought, Law and Culture*. Durham, NC: Duke University Press.

Dudley, M. J. (1996) Uncovering the invisible workers: using participatory vide and feminist methodologies to challenge distorted images of Colombian domestic workers. Masters Thesis, Cornell University.

Duneier, M. and Molotch, H. (1999) Talking city trouble: interactional vandalism, social inequality, and the 'urban interaction problem' *American Journal of Sociology* 5: 1263–95.

Durkheim, E. (1976) *The Elementary Forms of the Religious Life*. London: Allen & Unwin.

Dutton, P. (2004) Outside the Metropole: gentrification in provincial cities or provincial gentrification. In R. Atkinson and G. Bridge (eds), *Gentrification in a Global Context: The New Urban Colonialism*. London: Routledge.

Duvenage, P. (2003) *Habermas and Aesthetics: The Limits of Communicative Reason*. Cambridge: Polity.

Edmonson, M. (1956) Carnival in New Orleans. *Caribbean Quarterly* 4: 233–45.

Ehrenreich, B. and Ehrenreich, S. (1979) The professional-managerial class. In P. Walker (ed.), *Between Capital and Labour*. Sussex: The Harvester Press.

Ellison, R. (1964) *Shadow and Act*. New York: Random House.

Emirbayer, M. (1992) The shaping of a virtuous citizenry. *Studies in American Political Development* 6: 621–64.

Emirbayer, M. (1997) Manifesto for a relational sociology. *American Journal of Sociology* 103: 281–317.

Emirbayer, M. and Mische, A. (1998) What is agency? *American Journal of Sociology* 103: 962–1023.

Emirbayer, M. and Sheller. M. (1999) Publics in history. *Theory and Society* 28: 145–97.

Fincher, R. and Jacobs, J. (eds) (1998) *Cities of Difference*. New York: Guilford.

Fine, G. A. (ed.) (1995) *A Second Chicago School? The Development of Postwar American Sociology*. Chicago, IL: University of Chicago Press.

Flanagan, W. (1993) *Contemporary Urban Sociology*. Cambridge: Cambridge University Press.

Flyvbjerg, B. (1998) *Rationality and Power: Democracy in Practice*. Chicago, IL: University of Chicago Press.

Follett, M. P. (1965) *The New State: Group organisation the solution of popular government*. Gloucester, MA: Peter Smith (originally published in 1918).

Forester, J. (1989) *Planning in the Face of Power*. Berkeley: University of California Press.

Forester, J. (2000) *The Deliberative Practitioner*. London: MIT Press.

Forester, J. and Chu, L. (eds) (1990) Profiles of Planners (typescript). Working Papers 106 W. Sibley Hall, Cornell University, Ithaca, New York, 14853.

Foucault M. (1972) *The Archaeology of Knowledge* (trans. A. M. Sheridan Smith). New York: Harper and Row.

Foucault M. (1997) Ethics, subjectivity and truth. In P. Rabinow (ed.), *The Essential Works of Michel Foucault, 1954–1984*. New York: Free Press.

Foucault, M. (1977) *Discipline and Punish: The Birth of the Prison*. London: Allen Lane.

Foucault, M. (1978) *The History of Sexuality vol 1: An Introduction* (trans. R. Hurley). New York: Pantheon.

Foucault, M. (1986) Of other spaces. *Diacritics* 16: 22–7.

Fraser, N. (1992) Rethinking the public sphere: A contribution to the critique of actually existing democracy. In C. Calhoun (ed.), *Habermas and the Public Sphere*. Cambridge, MA: MIT Press, 109–42.

Friedmann, J. (1987) *Planning in the Public Domain: From Knowledge to Action*. Princeton, NJ: Princeton University Press.

Friedmann, J. (1992) *Empowerment: The Politics of Alternative Development*. Oxford: Blackwell.

Gagnon, J. and Simon, W. (1974) *Sexual Conduct: The Social Sources of Human Sexuality*. London: Hutchinson.

Gans, H. (1962) *The Urban Villagers: Group and Class in the Life of Italian Americans*. New York: Free Press.

Gardiner, M. (1996) Alterity and ethics: a dialogical perspective. *Theory, Culture and Society* 13: 121–43.

Garfinkel, H. (1967) *Studies in Ethnomethodology*. Englewood Cliffs, NJ: Prentice Hall.

Gatens, M. (1992) Power, bodies, difference. In M. Barratt and A. Phillips (eds), *Destabilising Theory*. Cambridge: Polity Press, 120–38.

Gibson, K. and Watson, S. (eds) (1995) *Postmodern Cities and Spaces*. Oxford: Blackwell.

Gilroy, P. (1993) *The Black Atlantic*. London: Verso.

Glass, R. (1964) *Introduction to London: Aspects of Change*. London: Centre for Urban Studies.

Goffman, E. (1955) On face work. *Psychiatry* 18: 213–231.

Goffman, E. (1959) *The Presentation of Self in Everyday Life*. Garden City, NY: Anchor.

Goffman, E. (1967) Where the action is. In E. Goffman (eds), *Interaction Ritual: Essays in Face-to-Face Behaviour*. New York: Pantheon, 149–270.

Goffman, E. (1970) *Strategic Interaction*. Oxford: Basil Blackwell.

Goffman, E. (1971) *Relations in Public*. London: Allen Lane, The Penguin Press.

Goffman, E. (1974) *Frame Analysis: An Essay on the Organisation of Experience*. Cambridge, MA: Harvard University Press.

Goffman, E. (1983) Felicity's condition. *American Journal of Sociology* 89: 1–53.

Gouldner, A. (1979) *The Future of Intellectuals and the Rise of the New Class: A Frame of Reference, Theses, Conjectures, Argumentation and a Historical Perspective*. London: Macmillan.

Granovetter, M. (1985) Economic action and social structure: the problem of embeddedness. *American Journal of Sociology* 91: 481–510.

Grosz, E. (1995) Bodies-Cities. In *Time, Space and Perversion*. London: Allen and Unwin, 103–10.

Habermas, J. (1984) *The Theory of Communicative Action, Volume 1: Reason and the Rationalisation of Society* (trans. by T. McCarthy). London: Heinemann.

Habermas, J. (1987) *The Theory of Communicative Action, Volume 2: A Critique of Functionalist Reason* (trans. by T. McCarthy). Cambridge: Polity.

Habermas, J. (1989) *The Structural Transformation of the Public Sphere* (trans. T. Burger). Cambridge, MA: MIT Press.

Habermas, J. (1990) *Moral Consciousness and Communicative Action* (trans. C. Lenhardt and S. W. Nicholsen). Cambridge, MA: MIT Press.

Hage, G. (1997) At home in the entrails of the west. In H. Grace *et al.* (eds), *Home/world: Space, Community and Marginality in Sydney's West*. Sydney: Pluto, 99–153.

Hall, P. (1988) *Cities of Tomorrow*. Oxford: Blackwell.

Hall, P. (1998) *Cities in Civilization*. London: Weidenfeld and Nicolson.

Hamnett, C. (1991) The blind men and the elephant: the explanation of gentrification. *Transactions of the Institute of British Geographers* 16: 173–89.

Hamnett, C. (1994) Social polarisation in global cities: theory and evidence. *Urban Studies* 31: 401–24.

Hannerz, U. (1990) Cosmopolitans and locals in a world culture. *Theory, Culture and Society* 7: 2–3.

Hannerz, U. (1996) *Transnational Connections: Culture, People, Places*. London: Comedia.

Harper, T. and Stein, S. (1996) Postmodernist planning theory: the incommensurability premise. In S. Mandelbaum, L. Mazza and R. Burchell (eds), *Explorations in Planning Theory*. New Brunswick, NJ: Centre for Urban Policy Research, 414–29.

Harris, J. (2001) Introduction. In F. Tonnies, *Community and Civil Society* (ed. J. Harris). Cambridge: Cambridge University Press (originally published 1887).

Harvey, D. (1973) *Social Justice and the City*. Baltimore, MD: Johns Hopkins University Press.

Harvey, D. (1978) The urban process under capitalism: a framework for analysis. *International Journal of Urban and Regional Research* 2: 101–31.

Harvey, D. (1985) *The Urbanisation of Capital*. Oxford: Blackwell.

Harvey, D. (1989) *The Condition of Postmodernity*. Oxford: Blackwell.

Harvey, D. (2000) *Spaces of Hope*. Edinburgh: Edinburgh University Press.

Hayden, D. (1980) What would a non-sexist city be like? In C. Stimpson *et al.* (eds), *Women and the American City*. Chicago, IL: University of Chicago Press.

Healey, P. (1997) *Collaborative Planning*. Basingstoke: Palgrave.

Healey, P. and Hillier, J. (1996) Communicative micropolitics. *International Planning Studies* 1.

Henderson, M. (1975) Acquiring privacy in public. *Urban Life and Culture* 3: 54–78.

Heskin, A (1991) *The Struggle for Community*. Boulder, CO: Westview Press.

Hicks, D. and Langsdorf, L. (1999) Regulating disagreement, constituting participants: a critique of proceduralist theories of democracy *Argumentation* 13: 139–60.

Hoch, C. (1984a) Doing good and being right: the pragmatic connection in planning theory. *Journal of the American Planning Association* 50: 335–45.

Hoch, C. (1984b) Pragmatism, planning and power. *Journal of Planning Education and Research* 4: 86–95.

Hoch, C. (1996) A pragmatic inquiry about planning and power. In S. Mandelbaum, L. Mazza and R. Burchell (eds). *Explorations in Planning Theory*. New Brunswick, NJ: Centre for Urban Policy Research, 30–44.

Hodgeson, G., Samuels, W. and Tool, M. (eds) (1993) *Companion to Institutional and Evolutionary Economics*. Aldershot: Edward Elgar.

Hoggett, P. and Thompson, S. (2002) Toward a democracy of the emotions. *Constellations* 9: 106–26.

Holston, J. (1989*)* *The Modernist City: An Anthropological Critique of Brasilia*. Chicago, IL: University of Chicago Press.

Honig, B. (1992) Toward an agonistic feminism. In J. Butler and J. Scott (eds), *Feminists Theorise the Political*. London: Routledge, 215–35.

Honneth, A. (1986) The fragmented world of symbolic forms: reflections on Pierre Bourdieu's sociology of culture. *Theory, Culture and Society* 3: 55–66.

Hoyt, H. (1939) *The Structure and Growth of Residential Neighborhoods in American Cities*. Washington, DC: Federal Housing Administration.

Huxley, M. (2000) The limits of communicative planning. *Journal of Planning Education and Research* 19: 369–77.

Huxley, M. and Yiftachel, O. (2000) New paradigm of old myopia? Unsettling the communicative turn in planning theory. *Journal of Planning Education and Research* 19: 333–42.

Innes, J. (1998) Information in communicative planning. *Journal of the American Planning Association* 64: 52–63.

Ishiguro, K. (1989) *The Remains of the Day*. New York: Vintage International.

Isin, E. (ed.) (2000) *Democracy, Citizenship and the Global Economy*. London: Routledge.

Jackson, P. and Smith, S. J. (1984) *Exploring Social Geography*. London: Allen and Unwin.

Jacobs, J. (1961) *The Life and Death of Great American Cities*. New York: Random House.

Jager, M. (1986) Class definition and the aesthetics of gentrification: Victoriana in Melbourne, Australia. In N. Smith and P. Williams (eds), *Gentrification of the City*. London: Allen and Unwin, 78–91.

Jameson, F. (1984) Postmodernism, or the cultural logic of late capitalism. *New Left Review* 146: 53–92.

Jankowiak, W. and White, C. (1999) Carnival on the clipboard: an ethnological study of New Orleans Mardi Gras. *Ethnology* 38: (4): 335–49.

Janowitz, M. (1967) *The Community Press in an Urban Setting*. Chicago, IL: University of Chicago Press.

Jenkins, R. (1992) *Pierre Bourdieu*. London: Routledge.

Joas, H. (1985) *G. H. Mead: A Contemporary Re-examination of his Thought* (trans. R. Meyer). Cambridge, MA: MIT Press.

Joas, H. (1993) *Pragmatism and Social Theory*. Chicago, IL: University of Chicago Press.

Johnson, J. (1991) Habermas on strategic and communicative action. *Political Theory* 19: 181–201.

Johnson, J. (1993) Is talk really cheap? Prompting conversation between critical theory and rational choice. *American Political Science Review* 87: 74–86.

Johnson, T. (1913) *My Story*. New York: B W Huebsch.

Jones, J. P. III (2000) The street politics of Jackie Smith. In G. Bridge and S. Watson (eds), *A Companion to the City*. Oxford: Blackwell, 448–59.

Judd, D. and Fainstein, S. (eds) (1999) *The Tourist City*. New Haven, CT: Yale University Press.

Kang, W. (1976) *G H Mead's Concept of Rationality: A Study in the Use of Symbols and other Implements*. The Hague: Mouton.

Karp, D. (1973) Hiding in pornographic bookstores. A reconsideration of the nature of urban anonymity. *Urban Life and Culture* 1: 427–51.

Kearns G. and Philo, C. (eds) (1993) *Selling Places: The City as Cultural Capital, Past and Present.* Oxford: Pergamon.

Kermond, C. (1996) Suffering the pain of separation. *The Age* (Melbourne), 19 Oct, A5.

King, M (1981) *Chain of Change.* Boston, MA: South End Press.

Knight, J. and Johnson, J. (1999) Inquiry into democracy: what might a pragmatist make of rational choice theory *American Journal of Political Science* 43: 566–89.

Kompridis, N. (1994) On world disclosure: Heidegger, Habermas, and Dewey. *Thesis Eleven* 37: 29–45.

Kotanyi, A. and Veneigem, R. (1961) Elementary programme of the board of unitary urbanism. *Internationale Socialiste* 6, and *Situationist International Anthology*, p. 67.

Kristeva, J. (1991) *Strangers to Ourselves* (trans. L. Roudiez). New York: Colombia University Press.

Kronman, A. (1987) *Living in the Law.* Chicago Law Review 54.

Laclau, E. and Mouffe, C. (1985) *Hegemony and Socialist Strategy: Towards a Radical Democratic Politics* (trans. W. Moore and P. Cammack). London: Verso.

Langsdorf, L. (1997) Argument as inquiry in a postmodern context *Argumentation* 11: 315–27.

Langsdorf, L. (2000) The possibility of communicative action. In L. E. Hahn (ed.), *Perspectives on Habermas.* Peru, IL: Open Court.

Langsdorf, L. (2002) Reconstructing the fourth dimension: A Deweyan critique of Habermas's conception of communicative action. In M. Aboulafia, M. Bookman and C. Kemp (eds), *Habermas and Pragmatism.* London: Routledge, 141–64.

Lash, S. (1999) *Another Modernity, A Different Rationality.* Oxford: Blackwell.

Laska, S. and Spain, D. (1980) *Back to the City: Issues in Neighborhood Renovation.* Elmsford, NY: Pergamon.

Latour, B. (1987) *Science in Action.* Milton Keynes: Open University Press.

Lauria, M. and Knopp, L. (1985) Toward an analysis of the role of gay communities in the urban renaissance. *Urban Geography* 6: 152–69.

Le Bon, G. (1966) *The Crowd.* New York: Viking Press (originally published in 1895).

Le Corbusier (1971) *The City of Tomorrow and its Planning* (3rd edn). London: Architectural Press.

Leavitt, J. (1994) Planning in an age of rebellion: guidelines to activist research and applied planning. *Planning Theory* 10: 111–30.

Leavitt, J. and Saegert, S. (1990) *From Abandonment to Hope: Community Households in Harlem.* New York: Colombia University Press.

Lefebvre, H. (1991) *The Production of Space.* Oxford: Blackwell.

Lefebvre, H. (1996) *Writings on Cities* (trans. E. Kofman and E. Lebas). Oxford: Blackwell.

Lewis, D. (1969) *Convention: A Philosophical Study.* Oxford: Basil Blackwell.

Ley, D. (1980) Liberal ideology and the postindustrial city. *Annals of the Association of American Geographers* 70: 238–58.

Ley, D. (1986) Alternative explanations for inner-city gentrification: a Canadian assessment. *Annals of the Association of American Geographers* 76: 521–35.

Ley, D. (1996) *The New Middle Class and the Remaking of the Central City.* Oxford: Oxford University Press.

Ley, D. (2003) Artists, aestheticisation and the field of gentrification. *Urban Studies* 40: 2527–43.

Lingis, A. (1998) *The Imperative*. Bloomington: Indiana University Press.

Lofland, J. (1976) *Doing Social Life: The Qualitative Study of Human Interaction in Everyday Life*. New York: Wiley.

Lofland, L. (1985) *A World of Strangers: Order and Action in Urban Public Space*. Prospect Heights, IL: Waveland.

Lofland, L. (1998) *The Public Realm: Exploring the City's Quintessential Social Territory*. New York: Aldine de Gruyter.

MacCannel, D. (1999) New Urbanism' and its discontents. In J. Copjec and M. Sorkin (eds), *Giving Ground: The Politics of Propinquity*. London: Verso, 106–30.

McCarthy, T. (1978) *The Critical Theory of Jurgen Habermas*. Cambridge, MA: MIT Press.

McIntosh, M. (1968) The homosexual role. *Social Problems* 16: 182–92.

McKenzie, E. (1996) *Privatopia: Homeowner Associations and the Rise of Residential Private Government*. New Haven, CT: Yale University Press.

McRobbie, A. (1994) *Postmodernism and Popular Culture*. London: Routledge.

McRobbie, A. (1984) Dance and social fantasy. In A. McRobbie and M. Nava (eds), *Gender and Generation*. London: Macmillan.

Maffesoli, M. (1996) *The Time of Tribes: The Decline of Individualism in Mass Society*. London: Sage.

Mattson, K. (1998) *Creating a Democratic Public: The Struggle for Urban Participatory Democracy during the Progressive Era*. University Park, PA: Penn State Press.

May, J. (1996) "A little taste of something more exotic": the imaginative geographies of everyday life. *Geography* 81: 57–64.

Mead, G. H. (1934) *Mind, Self and Society*. Chicago, IL: University of Chicago Press.

Mead, G. H. (1938) *The Philosophy of the Act*. C. Morris *et al.* (eds). Chicago, IL: Chicago University Press.

Mead, G. H. (1964) The social self. In A. Reck (ed.), *George Herbert Mead: Selected Writings*. Indianapolis, IN: Bobbs-Merrill, 142–49.

Merleau Ponty, M. (1962) *Phenomenology of Perception*. London: Routledge and Kegan Paul.

Mills, C. (1988) "Life on the upslope": the postmodern landscape of gentrification. *Environment and Planning D: Society and Space* 6: 169–89.

Mirowksi, P (1988) *Against Mechanism: Protecting Economics From Science*. Totowa, NJ: Rowman and Littlefield.

Mitchell, D. (1995) The end of public space? People's Park, definitions of the public and democracy. *Annals of the Association of American Geographers* 85: 108–33.

Mitchell, D. (2003) *The Right to the City: Social Justice and the Fight for Public Space*. New York: Guilford Press.

Moon, D. (1995) Practical discourse and communicative ethics. In S. K. White (ed.), *The Cambridge Companion to Habermas*. Cambridge: Cambridge University Press, 143–66.

Mort, F. (2000) The sexual geography of the city. In G. Bridge and S. Watson (eds), *A Companion to the City*. Oxford: Blackwell, 307–15.

Mouffe, C. (2000) *The Democratic Paradox*. London: Verso.

Ondaatje, M. (1992) *The English Patient*, New York: Vintage.

Park, R. E. (1926) The urban community as a spatial pattern and moral order. In E. W. Burgess (ed.), *The Urban Community*. Chicago, IL: University of Chicago Press,

Park, R. E. (1936) Human ecology. *American Journal of Sociology* 42: 1–15.

Park, R. E. (1969) The City: Suggestions for the Investigation of Human Behaviour in the Urban Environment. In R. Sennett (ed.), *Classic Essays in the Culture of Cities*. New York: Meredith (originally published in 1916).

Park, R. E. (1972) The crowd and the public. In *The Crowd and the Public and Other Essays*. Chicago, IL: Chicago University Press (originally published in 1904).

Park, R. E., Burgess, E. W. and McKenzie, R. D. (1925) *The City*. Chicago, IL: University of Chicago Press.

Pateman, C. (1970) *Participation and Democratic Theory*. Cambridge: Cambridge University Press.

Peach, C. (ed.) (1975) *Urban Social Segregation*. Harlow: Longman.

Peattie, L. (1994) Communities and interests in advocacy planning. *Journal of the American Planning Association* 60: 151–3.

Pile, S. (2000) Sleepwalking in the Modern City: Walter Benjamin and Sigmund Freud in the World of Dreams. In G. Bridge and S. Watson (eds), *A Companion to the City*. Oxford: Blackwell, 75–86.

Plant, S. (1992) *The Most Radical Gesture: The Situationist International in a Postmodern Age*. London: Routledge.

Plummer, K. (1975) *Sexual Stigma: An Interactionist Account*. London: Routledge and Kegan Paul.

Plummer, K. (ed.) (1997) *The Chicago School: Critical Assessments* (2 vols). London: Routledge.

Podmore, J. (1998) (Re)Reading the 'loft living' habitus in Montreal's inner city. *International Journal of Urban and Regional Research* 22: 283–302.

Pratt, G. (1998) Grids of difference: place and identity formation. In R. Fincher and J. Jacobs (eds), *Cities of Difference*. New York: Guilford, 26–48.

Rapaport, R. (1999) *Ritual and Religion in the Making of Humanity*. Cambridge: Cambridge University Press.

Redfern, P. (1997a) A new look at gentrification: 1. Gentrification and domestic technologies. *Environment and Planning A* 29: 1275–96.

Redfern, P. (1997b) A new look at gentrification: 2. A model of gentrification. *Environment and Planning A* 29: 1335–54.

Rhinegold, H. (2002) *Smart Mobs: The Next Social Revolution*. New York: Perseus Books.

Rittel, H. and Webber, M. (1973) Dilemmas in a general theory of planning. *Policy Sciences* 4: 155–69.

Robbins, B. (2001) The village of the liberal managerial class. In V. Dharwadker (ed.), *Cosmopolitan Geographies: New Locations in Literature and Culture*. London: Routledge, 15–32.

Robinson, W. J. (1914) My views on homosexuality. *American Journal of Urology* 10: 551.

Rorty, R. (1982) *Consequences of Pragmatism*. Minneapolis: University of Minnesota Press.

Rorty, R. (1988) *Contingency, Irony and Solidarity*. Cambridge: Cambridge University Press.

Rorty, R. (1991) *Objectivity, Relativism and Truth*. Cambridge: Cambridge University Press.

Rorty, R. (2000) Universality and truth. In R. Brandom (ed.), *Rorty and his Critics*. Oxford: Blackwell, 1–30.

Rose, D. (1984) Rethinking gentrification: beyond the uneven development of Marxist theory. *Environment and Planning D* 2: 47–74.

Rosenthal, S. (1990) *Speculative Pragmatism*. New York: Open Court.

Rosenthal, S. (2002) Habermas, Dewey and the Democratic Self. In M. Aboulafia, M. Bookman and C. Kemp (eds), *Habermas and Pragmatism*. London: Routledge, 210–22.

Rossiter, B. and Gibson, K. (2000) Walking and performing 'the City': A Melbourne chronicle. In G. Bridge and S. Watson (eds.), *A Companion to the City*. Oxford: Blackwell, 437–47.

Sandercock, L. (1998) *Towards Cosmopolis*, London: Wiley.

Sassen, S. (2000) *The Global City: New York, London and Tokyo* (updated edition). Princeton, NJ: Princeton University Press.

Schelling, T. (1980) *The Strategy of Conflict*. Cambridge, MA: Harvard University Press.

Schembri, J. (1996) The window of opportunities. *The Age* (Melbourne) 11 Oct: 5.

Schon, D. (1983) *The Reflective Practitioner*. New York: Basic Books.

Schrag, C. O. (1992) *The Resources of Rationality: A Response to the Postmodern Challenge*. Bloomington: Indiana University Press.

Scott, A. J. (1989) *Metropolis: From the Division of Labour to Urban Form*. Berkeley: University of California Press.

Searle, J. R. (1969) *Speech Acts: An Essay on the Philosophy of Language*. Cambridge: Cambridge University Press.

Sennett, R. (1970) *The Uses of Disorder*. Harmondsworth: Penguin.

Sennett, R. (1974) *The Fall of Public Man*. New York: Norton.

Sennett, R. (1990) *The Conscience of the Eye: The Design and Social Life of Cities*. London: Faber and Faber.

Sennett, R. (2000) Reflections on the public realm. In G. Bridge and S. Watson (eds), *A Companion to the City*. Oxford: Blackwell, 380–87.

Shalin, D. (1992) Critical theory and the pragmatist challenge. *American Journal of Sociology* 98: 237–79.

Shaw, K. (2004) Local limits to gentrification: Implications for a new urban policy. In R. Atkinson and G. Bridge (eds), *Gentrification in a Global Context: The New Urban Colonialism*. London: Routledge.

Shaw, W. (2004) Heritage and gentrification: remembering 'The Good Old Days' in postcolonial Sydney. In R. Atkinson and G. Bridge (eds), *Gentrification in a Global Context: the New Urban Colonialism*. London: Routledge.

Short, J. R. (1989) Yuppies, yuffies and the new urban order. *Transactions of the Institute of British Geographers* NS 14: 173–88.

Shott, S. (1976) Society, self, and mind in moral philosophy: the scottish moralists as precursors of symbolic interactionism. *Journal of the History of the Behavioural Sciences* 12: 30–46.

Shrum, W. and Kilburn, J. (1996) Ritual disrobement at Mardi Gras: ceremonial exchange and moral order. *Social Forces* 75: 423–58.

Siegfried, C. H. (1996) *Pragmatism and Feminism: Renewing the Social Fabric*. Chicago: University of Chicago Press.

Simmel G. (1950) The Metropolis and Mental Life. In K. Wolff (ed. and trans.), *The Sociology of Georg Simmel*. London: Free Press, 409–24 (originally published in 1903).

Simmel, G. (1990) *The Philosophy of Money*. In D. Frisby (ed. and trans T. Bottomore and D. Frisby). London: Routledge (originally published in 1907).

Simon, H. (1957) *Models of Man: Social and Rational*. New York: Wiley.

Smelser, N. and Swedberg, R. (eds) (1994) *The Handbook of Economic Sociology*. Princeton, NJ: Russell Sage Foundation.

Smith, N. (1979) Toward a theory of gentrification: a back to the city movement of capital not people. *Journal of the American Planning Association* 45: 538–48.

Smith, N. (1996) *The New Urban Frontier: Gentrification and the Revanchist City*. London: Routledge.

Smith, N. (2002) New globalism, new urbanism: gentrification as global urban strategy. *Antipode* 34(3): 428–50.

Soja, E. W. (1989) *Postmodern Geographies: The Reassertion of Space in Critical Social Theory*. Oxford: Blackwell.

Soja, E. W. (1996) *Thirdspace: Journeys into Los Angeles and Other Real and Imagined Places*. Oxford: Blackwell.

Soja, E. W. (1997) *Postmetropolis: Critical Studies of Cities and Regions*. Oxford: Blackwell.

Sorkin, M. (1992) *Variations on a Theme Park: The New American City and the End of Public Space*. New York: Hill and Wong.

Stein, M. (1960) *The Eclipse of Community*. Princeton, NJ: Princeton University Press.

Stuhr, J, (2003) *Pragmatism, Postmodernism, and the Future of Philosophy*. London: Routledge.

Sugden, R. (1989) Spontaneous order. *Journal of Economic Perspectives* 3: 85–97.

Sullivan, S. (2001) *Living Across and Through Skins: Transactional Bodies, Pragmatism, and Feminism*. Bloomington: Indiana University Press.

Suttles, G. (1968) *The Social Order of the Slum: Ethnicity and Territory in the Inner City*. Chicago, IL: University of Chicago Press.

Tajbakhsh, K. (2001) *The Promise of the City: Space, Identity, and Politics in Contemporary Social Thought*. London: University of California Press.

Throgmorton, J. (1996) *Planning as Persuasive Storytelling: The Rhetoric Construction of Chicago's Electric Future*. Chicago, IL: University of Chicago Press.

Tonnies, F. (2001) *Community and Civil Society* (ed. J. Harris). Cambridge: Cambridge University Press (originally published in 1887).

Ullman-Margalit, E. (1977) *The Emergence of Norms*. Oxford: Oxford University Press.

Urry, J. (2000) *Sociology Beyond Societies: Mobilities for the 21st Century*. London: Routledge.

Veblen, T. (1899) *The Theory of the Leisure Class*. New York: Macmillan.

Veblen, T. (1961) *The Place of Science in Modern Civilisation and Other Essays*. New York: B.W. Russell and Russell (originally published in 1919).

Verma, N. (1996) Pragmatic rationality and planning theory. *Journal of Planning Education and Research* 16: 5–14.

Verma, N. (1998) *Similarities, Connections and Systems: the Search for a New Rationality for Planning and Management*. London: Lexington Books.

Vidler, A. (2000) Photourbanism: planning the city from above and below. In G. Bridge and S. Watson (eds), *A Companion to the City*. Oxford: Blackwell, 35–45.

Watson, S. (1988) *Accommodating Inequality*. Sydney: Allen and Unwin.

Watson, S. (1999) A home is where the heart is: engendering notions of homelessness. In P. Kennett and A. Marsh (eds), *Homelessness: Exploring the New Terrain*. Bristol: Policy Press.

Webber, M. M. (1964) Urban place and the nonplace urban realm. In M. M. Webber *et al.* (eds), *Explorations into Urban Structure*. Philadelphia: University of Pennsylvania Press.

Weber, M. 1968 *Economy and Society* (3 vols). Totowa, NJ: Bedminster Press (originally published in 1921).

Wellmer, A. (1991) *The Persistence of Modernity: Essays on Aesthetics, Ethics and Postmodernism*. Cambridge: Polity.

Welsch, W. (1998) Rationality and reason today. In D. Gordon and J. Niznik (eds), *Criticism and Defense of Rationality in Contemporary Philosophy*. Amsterdam-Atlanta: Rodopi, 17–31.

West, C. (1993) *Race Matters*. Boston, MA: Beacon Press.

White, H. (1992) *Identity and Control: A Structural Theory of Social Action*. Princeton, NJ: Princeton University Press.

Willard, C. (1989) *A Theory of Argumentation*. London: University of Alabama Press.

Williamson, O. (1985) *The Economic Institutions of Capitalism: Firms, Markets and Relational Contracting*. London: Macmillan.

Wilson, E. (1991) *The Sphinx in the City: Urban Life, Control of Disorder, and Women*. Oxford: University of California Press.

Wirth, L. (1938) Urbanism as a way of life. *American Journal of Sociology* 44: 1–24.

Wolfinger, N. (1995) Passing moments. Some social dynamics of pedestrian interaction. *Journal of Contemporary Ethnography* 24: 323–40.

Wright, T. (1997) *Out of Place: Homeless Mobilisations, Subcities, and Contested Landscapes*. Albany: State University of New York Press.

Wyly, E. and Hammel, D. (2004) Mapping neoliberal American urbanism. In R. Atkinson and G. Bridge (eds), *Gentrification in a Global Context: The New Urban Colonialism*. London: Routledge.

Wynne, D. and O'Connor, J. (1998) Consumption and the postmodern city. *Urban Studies* 35: 841–64.

Young, I. M. (1990) The ideal of community and the politics of difference. In L. Nicholson (ed.) *Feminism/Postmodernism*. London: Verso, 300–23.

Young, I. M. (1997) Asymmetrical reciprocity: on moral respect, wonder and enlarged thought. *Constellations* 3: 340–63.

Young, I. M. (2000) *Inclusion and Democracy*. Oxford: Oxford University Press.

Younge, G. (2003) It's like Fight Club. *The Guardian* (1 August).

Zukin, S. (1982) *Loft Living: Culture and Capital in the Urban Change*. Baltimore, MD: Johns Hopkins University Press.

Zukin, S. (1991) *Landscapes of Power: From Detroit to Disney World*. Berkeley: University of California Press.

Zukin, S. (1995) *The Cultures of Cities*. Oxford: Blackwell.

Index